Rapid Methods for Assessing Food Safety and Quality

Rapid Methods for Assessing Food Safety and Quality

Editors

Maria Schirone
Pierina Visciano

MDPI • Basel • Beijing • Wuhan • Barcelona • Belgrade • Manchester • Tokyo • Cluj • Tianjin

Editors

Maria Schirone
University of Teramo
Italy

Pierina Visciano
University of Teramo
Italy

Editorial Office
MDPI
St. Alban-Anlage 66
4052 Basel, Switzerland

This is a reprint of articles from the Special Issue published online in the open access journal *Foods* (ISSN 2304-8158) (available at: https://www.mdpi.com/journal/foods/special_issues/Rapid_Methods_Assessing_Food_Safety_Quality).

For citation purposes, cite each article independently as indicated on the article page online and as indicated below:

LastName, A.A.; LastName, B.B.; LastName, C.C. Article Title. *Journal Name* **Year**, *Article Number, Page Range.*

ISBN 978-3-03943-344-5 (Hbk)
ISBN 978-3-03943-345-2 (PDF)

Contents

About the Editors

Maria Schirone Associate Professor of Inspection of Food of Animal Origin at the University of Teramo. Her research interests include: food safety with a particular attention to biological/chemical Hazards in food and beverages (biogenic amines, marine biotoxins, mycotoxins), physiological characterization and molecular typing of microbial ecosystems in fermented food production. Several studies have been conducted on the evaluation of the health standards of processes and/or products regarding some important food chains (e.g. dairy and fishery products, table olives, wine). She is also the author and co-author of many scientific publications edited in national and international journals, proceedings of national and international congresses, and book chapters. Moreover, she is the editor of the book "Igiene degli Alimenti—Aspetti igienico-sanitari degli alimenti di origine animale" (2014), Collana Edagricole Università & Formazione, and Member of the Scientific Committee. She participated as a speaker at Italian and International Conventions. She is a Guest Associate Editor for Frontiers in Microbiology and Topic Editor of two Research Topics on subjects associated with food safety (Biological Hazards in Food; Foodborne Pathogens: Hygiene and Safety) and a Research Topic for Fermentation (Microbial Control).

Pierina Visciano Associate Professor of Inspection of Food of Animal Origin at the University of Teramo. Co-author of many scientific publications edited in national and international peer-reviewed journals, book chapters, and proceedings of national and international congresses. Research activity is focused on food safety, biological hazards and hygiene requirements in the food industry and national and international food law. Specific expertise in chemical residues in food (veterinary drugs, polycyclic aromatic hydrocarbons, heavy metals, mycotoxins, marine biotoxins, histamine and biogenic amines), their reference doses relate to dietary intake, and maximum limits set by the European legislation. Further studies regard the evaluation of food safety and process hygiene criteria in regard to meat and fishery products, milk and dairy products. Guest Associate Editor of the international journal Frontiers in Microbiology. Topic Editor of two Research Topics on subjects associated with food safety (Biological Hazards in Food; Foodborne Pathogens: Hygiene and Safety). Editor of the book "Igiene degli Alimenti—Aspetti igienico-sanitari degli alimenti di origine animale" (2014), Collana Edagricole Università & Formazione, and Member of the Scientific Committee.

Preface to "Rapid Methods for Assessing Food Safety and Quality"

The quality and safety of foods are the crucial topic for both food industries and consumers, and their control represents an important objective to be guaranteed. The present Special Issue focuses on the development and application of rapid analytical methods able to identify the main failures in food production and to protect public health.

It consists of nine papers, Editorial included, on various food matrices analyzed by fast, cheap and reliable techniques, sometimes alternative to the recognized official methods. The use of quick and easy assays can benefit food business operators for good and reliable monitoring throughout the food chain.

Maria Schirone, Pierina Visciano

Editors

Editorial

Rapid Methods for Assessing Food Safety and Quality

Pierina Visciano and Maria Schirone *

Faculty of Bioscience and Technology for Food, Agriculture and Environment, University of Teramo,
64100 Teramo, Italy; pvisciano@unite.it
* Correspondence: mschirone@unite.it; Tel.: +39-0861-266911

Received: 14 April 2020; Accepted: 16 April 2020; Published: 23 April 2020

Abstract: Food safety represents a central issue for the global food chain and a daily concern for all people. Contaminated food by physical, biological or chemical hazards can harm consumers, increasing demand for health services, government expenditure on public health and other social costs. The quality assurance programs are based on the continuous monitoring of raw matter, production process, storage and distribution of the end products, including the purpose for which they are intended. Such programs represent an important objective for food producers, not only for the potential risk to human health, but also for the economic losses to which they can be subjected. The development and use of rapid analytical methods able to identify the main failures in food production can benefit food companies by saving time and costs for the good and fast control of products through the entire food chain.

Keywords: safety; assay; pollutants; polyciclic aromatic hydrocarbons; *Trichinella*; *Anisakis*; *Listeria monocytogenes*; nanoparticles

---

Nowadays, food safety has a critical societal importance for producers, regulatory control bodies and consumers. For this purpose, the food industry requires fast, sensitive, reliable, cost-effective and easy-to-use analytical techniques to assess both the safety and quality of products [1,2]. In the last few years, the traditional culture-based enumeration tests used for the detection of microorganisms have become obsolete for real-time applications, as they are time-consuming and labor-intensive, while immunological assays such as the enzyme immunoassay, enzyme-linked immunosorbent assay, enzyme-linked fluorescent assay, flow injection immunoassay and other serological methods are known for their speed, high-throughput capacity and possibility of precise quantification of the target organism [3].

Even for chemical analysis, the development of quick and easy methods is beginning to be preferred to classical laborious techniques [4]. One of these pre-treatment assays for sample preparation is based on the QuEChERS method, with the advantages summarized in its acronym (quick, easy, cheap, effective, rugged, and safe). The study of Nagyová and Tölgyessy [5] reported the validation of a rapid and non-laborious method for the determination of selected H_2SO_4 stable halogenated priority pollutants (eight organochlorines and six polybrominated diphenyl ethers) in nine different fish species. A modified QuEChERS sample preparation followed by gas chromatography-triple quadrupole tandem mass spectrometry analysis was described. In particular, the applied method showed some advantages in terms of simplicity, rapidity, high extract clean-up efficiency and good sensitivity. Moreover, the proposed assay can be classified as "an acceptable green analysis method" with low consumption of hazardous solvents.

Puljić et al. [6] evaluated the differences in polyciclic aromatic hydrocarbons (PAHs) content in samples of traditional dry cured pork meat products made in Herzegovina, subjected to uncontrolled (traditional smokehouse) and industrial smoking processes. The results highlighted that the use of traditional smoking methods resulted in higher PAH contamination than the industrial one. At the end of production, the inner parts of all smoked samples produced using both methods retained significantly lower total PAH concentration, as well as less individual PAHs than the surface layer. The amount of the four priority PAHs in samples subjected to traditional smoking highly exceeded maximum limits (12 µg/kg) set by the Commission Regulation (EU) No 835/2011 [7] by up to 10 times. The results of this study indicate that, in order to decrease the level of PAHs and reduce the risk of PAH occurrence in smoked meat products, local producers should learn how to use the improved/novel smoking techniques and adjust the smoking parameters.

Chen [8] studied the correlation between water activity (Aw) and moisture content (MC) of floral honey at 10 and 30 °C using the Aw method. It was observed that the temperature significantly affected the Aw/MC data, but no universal linear equation for these parameters could be established. In this study, the slope was affected only by the state of the honey (liquid or crystallized), while the other factors, such as type, geographical collection sites and botanical source did not significantly affect the slope, but only the intercept of the Aw equation. The results demonstrate that such a linear equation could be used to express the relationship between Aw and MC of honeys.

Among the food biological hazards, the presence of parasites is disgusting and dangerous at the same time. The paper by Schirone et al. [9] described the accreditation procedure and the requirements for its maintenance applied by an internal laboratory attached to a slaughterhouse for the detection of *Trichinella* spp. in swine carcasses. The main advantages were represented by the possibility to analyze the muscle samples quickly, in order to obtain fast results and process carcasses after a short time from slaughtering. On the other hand, the main difficulties shown by the technician working in the internal laboratory were the structural features required by the accreditation body, i.e., the proficiency testing to be carried out each year and the training courses to be followed in order to maintain the accredited status. However, the possibility of performing the regulatory investigation of *Trichinella* spp. in swine and other species susceptible to the parasite inside the slaughterhouse is useful for customers as well as for the competent authority in guaranteeing the safety of the slaughtered carcasses.

The study of Cammilleri et al. [10] described a rapid, reliable and easy assay for the detection of *Anisakis* spp. Such a method based on loop-mediated isothermal amplification (LAMP) was optimized, validated and applied in processed fish samples artificially contaminated with the parasite, giving sensitivity values equal to 100%. Indeed, the specificity test provided no amplification for other similar genera of parasites, i.e., *Contracaecum*, *Pseudoterranova*, or *Hysterothylacium*, as well as for uninfected samples. Moreover, the LAMP assay showed both a lower limit of detection and time of analysis in comparison with the real-time PCR method also used in this study, and its ease of use suggested that it could be a valid alternative for routine examination to help manufacturers in HACCP (Hazard Analysis Critical Control Point) application in the fishery sector.

Among the key attributes of a microbial detection system, there are sensitivity, i.e., the ability to find the target microorganism at low contamination level, and specificity, which allows the selective detection of the target microorganism without cross-reaction with other species. However, rapidity is also a good feature of an analytical assay to obtain fast and reliable results [11]. The nucleic acid-based techniques aim to detect the specific nucleic acid sequence by amplifying the target sequence to millions of folds. They usually provide more timely and accurate results compared with traditional immunoassays and culturing methods [12]. The study of Torresi et al. [13] applied a specific real-time PCR screening assay designed for the rapid identification of *Listeria monocytogenes* strains potentially related to the outbreaks occurred in Central Italy from January 2015 to March 2016. A total of 37 clinical strains isolated from patients exhibiting listeriosis symptoms and 1374 strains correlated to the outbreak were examined by the Italian National Reference Laboratory for this microorganism with a first screening assay using a specific real-time PCR. The rapid applied method was able to quickly screen up to 96 strains in 2–4 h and to identify all strains linked to the outbreak. Then, the PCR-positive strains were subsequently typed by other tests and precisely pulsed field gel electrophoresis and next generation sequencing. The results highlight the decrease in the time and costs of analysis, as well as the importance of the rapid identification of contaminated food and food processing industries to stop the diffusion of pathogenic strains and minimize the number of cases in a foodborne outbreak.

In recent decades, novel rapid and non-destructive methods for evaluating food quality attributes have been developed. In particular, new detection assays have been explored such as biosensors and novel nanotechnology-based immunoassay are quickly replacing traditional methods offering some advantages to food business operators. Unlike conventional methods, these techniques acquire data without contact with samples, and some examples are computer vision, spectral or hyperspectral imaging, ultrasound, near-infrared spectroscopy (NIR), Fourier transform near infrared spectroscopy (FT-NIR), and Raman spectroscopy [14]. Basile et al. [15] used a fast, simple and non-destructive technique, namely FT-NIR spectroscopy combined with chemometrics, applied to search for a correlation between the chemical composition in terms of sugar and acidic content and sensory data. Two very different table grape varieties were investigated by chemical (total soluble solid content, titratable acidity, and HPLC analysis of organic acids), FT-NIR and sensory analyses. The results provide a good basis for more elaborate studies to correlate sensory properties and chemical composition of food.

Recently, the use of nanosensors showed the advantages of high efficiency, rapidity and sensitivity, and therefore they were applied for clinical diagnosis, pesticide residue analysis and microbiological assays. Bi et al. [16] reported a study about the detection of histamine, a spoilage monitoring compound for distinguishing the lifetime and freshness of fishery products, by using gold nanoparticles (Au-NPs) with dual approaches of colorimetric and fluorescence perspectives. The Fourier transforms infrared (FTIR) spectra provided histamine concentrations in spiked salmon muscle samples with recovery values ranging from 96 to 103%. Besides high sensitivity and selectivity, such a system was characterized by the simple pre-treatment of samples, suitable equipment requirements and low costs, so that it could be used for freshness and spoilage determination of fish, as well as for safety assurance purposes.

To conclude, the present Special Issue consists of eight papers on various food matrices analyzed by fast, cheap and reliable techniques, sometimes as an alternative to the recognized official methods. Most of them aimed at verifying food safety as a fundamental attribute for the food processing industry, food retailers and distributors, and competent authorities due to the potential direct impact on consumer health.

Author Contributions: The authors have made a substantial, direct and intellectual contribution to the work and approved it for publication. All authors have read and agreed to the published version of the manuscript.

Funding: This research received no external funding.

Conflicts of Interest: The authors declare no conflict of interest.

References

1. Poghossian, A.; Geissler, H.; Schöning, M.J. Rapid methods and sensors for milk quality monitoring and spoilage detection. *Biosens. Bioelectron.* **2019**, *140*, 111272. [CrossRef] [PubMed]
2. Ziyaina, M.; Rasco, B.; Sablani, S.S. Rapid methods of microbial detection in dairy products. *Food Control* **2020**, *110*, 107008. [CrossRef]
3. Jayan, H.; Pu, H.; Sun, D.-W. Recent development in rapid detection techniques for microorganism activities in food matrices using bio-recognition: A review. *Trends Food Sci. Technol.* **2020**, *95*, 233–246. [CrossRef]
4. Musarurwa, H.; Chimuka, L.; Pakade, V.E.; Tavengwa, N.T. Recent developments and applications of QuEChERS based tecniques on food samples during pesticide analysis. *J. Food Compost. Anal.* **2019**, *84*, 103314. [CrossRef]
5. Nagyová, S.; Tölgyessy, P. Validation including uncertainty estimation of a GC-MS/MS method for determination of selected halogenated priority substances in fish using rapid and efficient lipid removing sample preparation. *Foods* **2019**, *8*, 101. [CrossRef]
6. Puljić, L.; Mastanjević, K.; Kartalović, B.; Kovačević, D.; Vranešević, J.; Mastanjević, K. The influence of different smoking procedures on the content of 16 PAHs in traditional dry cured smoked meat "Hercegovačka pečenica". *Foods* **2019**, *8*, 690. [CrossRef]

7. The European Commission. Commission Regulation (EU) No 835/2011 of 19 August 2011 amending Regulation (EC) No 1881/2006 as regards maximum levels for polycyclic aromatic hydrocarbons in foodstuffs. *Off. J. Eur. Union* **2011**, *50*, 2011.
8. Chen, C. Relationship between water activity and moisture content in floral honey. *Foods* **2019**, *8*, 30. [CrossRef]
9. Schirone, M.; Visciano, P.; Olivastri, A.M.A.; Sgalippa, M.P.; Paparella, A. Accreditation procedure for *Trichinella* spp. detection in slaughterhouses: The experience of an internal laboratory in Italy. *Foods* **2019**, *8*, 195. [CrossRef] [PubMed]
10. Cammilleri, G.; Ferrantelli, V.; Pulvirenti, A.; Drago, C.; Stampone, G.; Del Rocio Quintero Macias, G.; Drago, S.; Arcoleo, G.; Costa, A.; Geraci, F.; et al. Validation of a commercial loop-mediated isothermal amplification (LAMP) assay for the rapid detection of *Anisakis* spp. DNA in processed fish products. *Foods* **2020**, *9*, 92. [CrossRef]
11. Osopale, B.A.; Adewumi, G.A.; Witthuhn, R.C.; Kuloyo, O.O.; Oguntoyinbo, F.A. A review of innovative tecniques for rapid detection and enrichment of Alicyclobacillus durin industrial processing of fruit juices and concentrates. *Food Control* **2019**, *99*, 146–157. [CrossRef]
12. Zhang, M.; Ye, J.; He, J.-S.; Zhang, F.; Ping, J.; Qian, C. Visual detection for nucleic acid-based techniques as potential on-site detection methods. A review. *Anal. Chim. Acta* **2020**, *1099*, 1–15. [CrossRef] [PubMed]
13. Torresi, M.; Ruolo, A.; Acciari, V.A.; Ancora, M.; Blasi, G.; Cammà, C.; Centorame, P.; Centorotola, G.; Curini, V.; Guidi, F.; et al. A real-time PCR screening assay for rapid detection of *Listeria monocytogenes* outbreak strains. *Foods* **2020**, *9*, 67. [CrossRef] [PubMed]
14. Lei, T.; Sun, D.-W. Developments of nondestructive techniques for evaluating quality attributes of cheeses: A review. *Trends Food Sci. Technol.* **2019**, *88*, 527–542. [CrossRef]
15. Basile, T.; Marsico, A.D.; Cardone, M.F.; Antonacci, D.; Perniola, R. FT-NIR analysis of intact table grape berries to understand consumer preference driving factors. *Foods* **2020**, *9*, 98. [CrossRef] [PubMed]
16. Bi, J.; Tian, C.; Zhang, G.-L.; Hao, H.; Hou, H.-M. Detection of histamine based on gold nanoparticles with dual sensor system of colorimetric and fluorescence. *Foods* **2020**, *9*, 316. [CrossRef] [PubMed]

Article

Detection of Histamine Based on Gold Nanoparticles with Dual Sensor System of Colorimetric and Fluorescence

Jingran Bi [1,2,*], Chuan Tian [1,2], Gong-Liang Zhang [1,2], Hongshun Hao [1,2] and Hong-Man Hou [1,2,*]

[1] School of Food Science and Technology, Dalian Polytechnic University, No. 1, Qinggongyuan, Ganjingzi District, Dalian 116034, Liaoning, China; 18900986803@163.com (C.T.); zgl_mp@163.com (G.-L.Z.); beike1952@163.com (H.H.)

[2] Liaoning Key Lab for Aquatic Processing Quality and Safety, No. 1, Qinggongyuan, Ganjingzi District, Dalian 116034, Liaoning, China

* Correspondence: bijingran1225@foxmail.com (J.B.); houhongman@dlpu.edu.cn (H.-M.H.); Tel.: +86-411-8632-2020 (J.B.)

Received: 15 February 2020; Accepted: 5 March 2020; Published: 9 March 2020

Abstract: Gold nanoparticles (Au-NPs), with the dual sensor system of colorimetric and fluorescence responses, were developed for the determination of histamine as a spoilage monitor for distinguishing lifetime and freshness of aquatic products. Upon addition of histamine, the absorption coefficient orders of magnitude via the interaction of free electrons and photons were affected, and the characteristic absorption peak of Au-NPs was red-shifted from 520 nm to 664 nm. Meanwhile, the large amino groups in the networks of histamine-Au-NPs with high molecular orbital exhibited excellent fluorescence behavior at 415 nm. Au-NPs offered a range of 0.001–10.0 μM and 0.01–1.0 μM with a limit of detection of 0.87 nM and 2.04 nM by UV-vis and fluorescence spectrum assay, respectively. Moreover, Au-NPs could be used to semiquantitatively analyze histamine with the naked eye, since the significant colorimetric and fluorescence reaction of Au-NPs solution that coincided with different concentrations of histamine can be observed as the histamine concentration was 0.1–1.0 μM. Both of the dual-sensor systems of Au-NPs were successfully applied to the quantitative analysis of histamine in fresh salmon muscle, suggesting the simplicity and rapidity in the dual detection approaches of Au-NPs might be suitable for spoilage assay of aquatic food to ensure food safety.

Keywords: gold nanoparticles; histamine; UV-visible and fluorescence; visual detection; spoilage marker

1. Introduction

Biogenic amines are a type of significant toxic substances directly involved in food quality and safety. Under improper storage conditions, it is easy to cause spoilage bacteria to grow and metabolize decarboxylase, resulting in decarboxylation of amino acids in high-protein foods (such as dairy products, seafood, meat products, etc.) to form various biological amines, mainly including histamine, tyramine, cadaverine, and putrescine [1]. In particular, histamine (4-(2-aminoethyl)-1H-imidazole) is one of the most important biogenic amines, commonly formed by free L-histidine under the action of exogenous decarboxylase produced by microbial metabolism [2]. Once formed, histamine is hard to degrade by conventional methods such as heating and freezing due to its excellent stability. Moreover, food-borne histamine as an alkaline nitrogen-containing hazardous substance can affect the respiratory system and digestive system and frequently causes food poisoning around the world [3]. Hence, many countries have established the strict guidelines of histamine to prevent histamine food poisoning. The US Food and Drug Administration (FDA) set the maximum thresholds of histamine as 500 mg/kg [4]. The European Union 2073/2005 considers the histamine limits of 100–200 mg/kg for the fishery products from fish species as safe [5]. Due to the fact that histamine level generally increases as the improper storage time extends, histamine has become a globally recognized monitor for distinguishing lifetime and freshness of aquatic products [6]. Consequently, a direct and rapid sensor for detecting histamine in food is sorely required to guarantee food safety.

For decades, various standardized laboratory analytical techniques have been applied for the determination of histamine, such as high-performance liquid chromatography (HPLC) [7], thin-layer chromatography [8], and gas chromatography–tandem mass spectrometry analysis [9]. Although these techniques are to detect histamine in biomatrix, unfortunately, they require too much laborious sample preparations, instruction expertise operations, and time-consuming procedures to be practically applied in fishery enterprises and market supervision. Hence, the establishment of a simple ultrasensitive method for histamine detection is significant and challenging.

Recently, nanosensors with the advantages of high efficiency, rapidity, and sensitivity gradually caused great concern in many areas, such as clinical diagnosis [10], pesticide residue analysis [11], and microbiological detection [12]. Due to the food industry inextricably linked to public health, more and more attention has focused on nanosensors for the determination of food freshness to guarantee public health safety [13].

Although histamine does not absorb strongly, it can be detected by the UV-vis spectrum after derivatization transforms or chemosensory combination. Numerous receptors have been successfully designed to detect biomolecules by hydrogen bonds [14,15]. However, water molecules typically compete with analytes for binding to receptors in aqueous solutions [16]. Hence, metal ions are regarded as an appropriate probe for forming complexes with biomolecules by electrostatic interaction and are suitable for bioanalysis of samples with high moisture content.

Gold nanoparticles (Au-NPs) have unique dispersibility, long-term stability, sensitive optical properties, and facile surface functionalization, and have been widely applied in biosensing determination, such as the detection of toxin [17,18], DNA [19], and amino acid [20]. However, up to now, Au-NPs probes for detecting molecular species have been based almost exclusively on colorimetric properties, lack multisensor properties, and cannot be flexibly applied in practical analysis.

In this work, by the precursor ratio, an economical and practical new Au-NPs biomarker was fabricated for the detection of histamine with dual-sensor analysis. Due to the electrostatic interaction between histamine and Au-NPs, which increases the electron density and surface amino group content, histamine can be detected by Au-NPs via the dual approaches of colorimetric and fluoroscopy with high sensitivity, selectivity, and efficiency. Moreover, Au-NPs can be used to directly quantify whether the histamine content of fresh salmon muscle meets FDA standards, in the hope that this probe will be able to develop assays for the determination of freshness and spoilage of aquatic foods to ensure food safety.

2. Materials and Methods

2.1. Materials and Reagents

Chloroauric acid tetrahydrate (HAuCl$_4$·4H$_2$O) was purchased from Sangon Biotech (Shanghai, China). Histamine dihydrochloride, histidine, putrescine dihydrochloride, cadaverine dihydrochloride, 2-Phenylethylamine hydrochloride, and tyramine hydrochloride were obtained from Sigma (St. Louis, MO, USA) (http://www.sigmaaldrich.com). Trisodium citrate and NaCl were purchased from Tianjin Chemical Reagent Co., Ltd. (Tianjin, China). Salmon samples were purchased from a local aquatic products market. Moreover, Milli-Q-purified water (18.2 MΩcm) was used throughout the experiments.

2.2. Instrumentation

The morphology analysis was measured by the transmission electron microscope (JEM-2100, JEOL, Tokyo, Japan). UV-vis absorption spectra were obtained at room temperature via a UV-vis spectrophotometer (Lambda 35, PerkinElmer, Cambridge, MA, USA). Fluorescence spectra were studied by a fluorescence spectrometer (F-2700, Hitachi, Tokyo, Japan). Moreover, Fourier transforms infrared (FTIR) spectra were recorded at room temperature on the Frontier FTIR spectrometer (PerkinElmer, Norwalk, CT, USA). Average hydrodynamic diameter and polydispersity index (PDI) were measured by Dynamic Light Scattering (DLS) using the particle size analyzer (Zetasizer 3000HSA, Malvern, UK). Surface electric properties were analyzed by zeta potential analyzer (NanoZS ZEN3600, Malvern, UK).

2.3. Preparation of Citrate-Capped Au-NPs

Every piece of glassware was soaked in an aqua regia solution and completely washed by purified water before use. Two milliliters of 0.85% $HAuCl_4$ solution was cautiously added in 48 mL boiling water, dropwise, until the final concentration was around 1.0 mM. After stirring for 5 min, Trisodium citrate was quickly dropped into the boiled $HAuCl_4$ solution with different final precursor ratios of $HAuCl_4/Na_3Ct$ (1:2, 1:4, 1:6, and 1:8), followed by continuous stirring and heating for 15 min. Consequently, the color of the solution gradually changed to wine red, demonstrating that Au-NPs were successfully synthesized by reducing Au (III) to Au (0) via trisodium citrate. Then, the Au-NPs solution was cooled and stored at 4 °C for reserve.

2.4. Detection of Histamine

Histamine was dissolved into 0.9% NaCl solution to achieve various final concentrations and added into Au-NPs solution ($v/v = 1/1$) with vigorous vortex-mixing for 2 min. Then, UV-vis spectra and fluorescence spectra were measured for detecting assay. The selectivity of the Au-NPs probe was also investigated by using different biological substances containing amino groups (histamine, putrescine, cadaverine, tyramine, phenylethylamine, guanine, guanosine, thymine, inosine, adenosine triphosphate (ATP), and adenosine monophosphate (AMP)).

2.5. Detection of Histamine In Salmon Muscle

The fresh salmon muscle was added into 0.9% NaCl solution ($w/v = 1/10$) and high-speed homogenate for 2 min. After being centrifuged at 4000 rpm at 4 °C for 10 min, the supernate was spiked with histamine (various final concentrations of 0.01, 0.1, 1.0 μM) and stirred thoroughly. Then, the samples were mixed with Au-NPs solution at $v/v = 1/1$ and measured by UV-vis spectra and fluorescence spectra to detect the concentration of histamine. Recovery was calculated from found concentration/known concentration × 100%. Precision was calculated from standard deviation/mean × 100%. Accuracy was calculated from (found concentration − known concentration)/known concentration × 100%.

2.6. Statistical Analysis

All measurements in the study were done in triplicate. The results reported here were the means of the three trials. The results were expressed as means ± standard deviation (SD). All the diagrams were plotted by Origin 9.2 software (Microcal, Northampton, MA, USA).

3. Results and Discussion

Au-NPs were successfully synthesized by being reduced and capped via citrate in boiling water (Scheme S1) with $HAuCl_4/Na_3Ct$ precursor ratio of 1:6 (Figure S1). Au-NPs had extremely small particle size, good monodispersity, and strong UV-vis absorption at 520nm, indicating that colloidal gold with intense surface plasmon resonance absorption at visible wavelengths was formed, which can be further analyzed as a sensor to detect histamine.

3.1. Sensing Mechanism of Au-NPs Probe

To explore the sensing mechanism of histamine detected by Au-NPs, FTIR spectrum was applied to study the structural changes, and the results illuminated the distinct functional groups of the reaction system. As shown in Figure 1, the correlation absorption peaks of Au-NPs at 1637 cm^{-1}

and 1388 cm^{-1} are attributed to $\sigma_{C=O}$ and σ_{C-O}, respectively, due to the decarboxylation molecules of synthetic product. Also, the strong characteristic band of histamine in the range of 1650–1600 cm^{-1} is assigned to C=N, and the absorption peak at 1564 cm^{-1} belongs to the in-plane bending vibration of the amine group. The band of 1120–1030 cm^{-1} corresponds to σ_{C-N}, and the primary amine groups were observed at 850–650 cm^{-1} and attributed to γ_{N-H}. After reacting with Au-NPs, the position of amine groups, including δ_{C-N}, σ_{C-N}, and γ_{N-H}, were remarkably strengthened; furthermore, the band of C=N double bond was also significantly enhanced, suggesting the positively charged functional groups might participate in electrostatic interactions with Au-NPs and large amine groups and unsaturated bonds on the surface of histamine-Au-NPs.

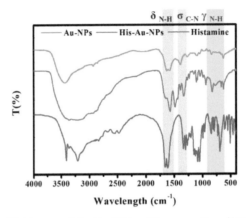

Figure 1. FTIR of Gold nanoparticles (Au-NPs) and the aggregation of histamine-Au-NPs.

By DLS measure (Figure 2A,B), the average hydrodynamic diameter of Au-NPs after reaction with histamine increased from 1.3 nm to 290.2 nm, with a PDI value of 0.089 to 0.272, indicating that after the gold nanoparticles were combined with histamine, the dispersity was reduced, and the aggregation was enhanced. *Zeta* potential is consistent with DLS results, and the surface of Au-NPs is negative in charge (−28.31 ± 4.35 mV), while the histamine–Au-NPs surface is positive in charge (35.70 ± 2.96 mV) due to the NH$_2$ functional group in the histamine, which enables it to attach to the negative charge of AuNPs. TEM micrograph (Figure 2C,D) also confirmed the active surface affinity of Au-NPs towards histamine. Compared with Au-NPs, the morphology of histamine-Au-NPs appears significantly aggregated.

The key to the observed remarkable chromogenic reaction and fluorescence response is believed to be an amino role (Scheme 1). Au-NPs are negatively charged by the synthesis of Au (III) reduced and capped via citrate, and the electrostatic interaction occurred when Au-NPs are attracted by histamine that narrowed the interparticle distance between them. Histamine–Au-NPs are formed into massive networks, and the absorption coefficient orders of magnitude via the interaction of free-electrons and photons might be severely affected surface plasmon resonance absorptions enhance [21]. Moreover, the amino groups are linked with the Au-NPs, and n-π^* conjugate occurs. Hence, the UV-vis absorption was distinctly red-shifted from wine red to dark blue, which could be observed by eye vision. Meanwhile, the large amino groups in the networks of histamine-Au-NPs have higher molecular orbital (LUMO orbital) in contrast to the original hydrogen-terminated groups of Au-NPs. As the electrons are excited, they would be relaxed to the ground state through a narrow optical band gap with the amino-fluorophore vibrations/rotations [22]. Hence, the solution of histamine-Au-NPs could exhibit excellent turn-on fluorescence behavior.

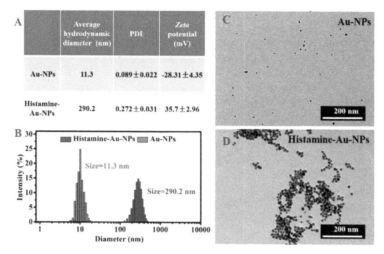

Figure 2. (**A**) Average hydrodynamic diameter, polydispersity index (PDI) and zeta potential information, (**B**) Dynamic Light Scattering (DLS), and (**C–D**) TEM images of Au-NPs and the aggregation of histamine-Au-NPs.

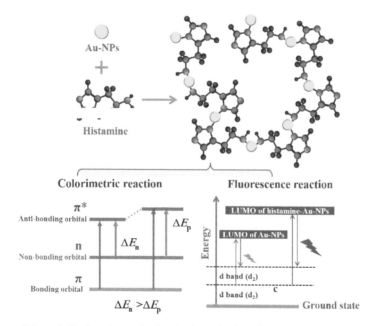

Scheme 1. Dual sensing mechanism for determination of histamine by Au-NPs.

3.2. Analytical Performance of Au-NPs Probe

Compared to the wine-red Au-NPs, on the inclusion of histamine, the color of the complex solution gradually turned into dark blue, with the increased concentrations (0.001–10.0 μM) as shown in Figure 3B, that was evident enough to semiquantitatively obtain the existence of histamine by visual observation. The multicolored appearance of metallic suspension would be ascribed to their surface plasmon polaritons. The extremely tiny NPs can combine with the analyte, dramatically influencing their optical characteristics; that is to say, metallic NPs can transform the color by the change of size and dispersity. Similarly, after the gold nanoparticles reacted with histamine, the monodispersity was significantly reduced and the surface plasmon polaritons was affected.. Histamine can form a rough layer on the surface of Au-NPs through electrostatic action. The roughness of Au-NPs causes the scattering of high-momentum surface plasmons, which can be coupled with radiated light by the loss of their momentum. This process is much faster than the spontaneous recombination of exciton dipoles, so the radiation intensity enhanced [23]. In addition, the UV-vis spectrums of various concentrations of histamine detected by Au-NPs were measured, as shown in Figure 3A. The characteristic peak of Au-NPs at 520 nm was gradually weakened, and a new peak at 664 nm was progressively enhanced with the increasing concentrations of histamine from 0.001 to 10.0 μM due to the changes of electron density by complexation between histamine and Au-NPs. Furthermore, there was a tremendous linear relationship between the absorption ratios (A_{664}/A_{520}) and the logarithm value of histamine concentrations in the dynamic range from 0.001 to 10.0 μM with the regression equation $y = 1.03815 + 0.10408x$ (Figure 3C). The correlation coefficient of the calibration curve was 0.9986. The detection limit (DL = $3S_{b1}/S$, where S_{b1} is the standard deviation of the blank solution and S is the slope of the calibration curve) was 0.87 nM.

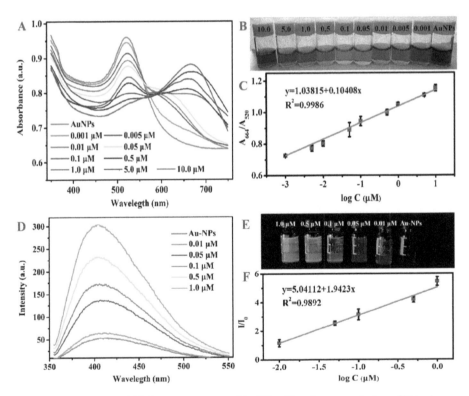

Figure 3. (**A**) UV–vis and (**D**) fluorescence spectra of Au-NPs with various concentrations of histamine. The corresponding (**B**) colorimetric and (**E**) fluorescent response of histamine-Au-NPs under daylight and 365nm UV-light. Linear relationships between (**C**) the absorption ratios (A_{664}/A_{520}), (**F**) the fluorescence intensity ratios (I/I_0), and the logarithm value of histamine concentrations.

Furthermore, by successive addition of histamine, the electrostatic interacted amino groups were progressively increased, the glaucous fluorescence of histamine-Au-NPs solution exhibited brighter and brighter under UV lamp (λ = 365 nm) (Figure 3E). As the concentration of histamine became higher than 0.1 μM, the significant colorimetric and fluorescence turn-on response with Au-NPs could be captured by eye-vision. As shown in Figure 3D, histamine could also be detected by Au-NPs via fluorescence spectra analysis. The emission peak around 415 nm was strengthened gradually as the concentrations of histamine continuously increased in the range of 0.01–1.0 μM with an excellent linear relationship with the regression equation y = 5.04112x + 1.9423 (where y is the fluorescence intensity ratio between before and after adding histamine and x is the logarithm value of histamine concentrations) (Figure 3F). The correlation coefficient of the calibration curve was 0.9986. Moreover, the Au-NPs probe for detecting histamine performed a limit of detection as 2.04 nM.

Previously, HPLC was considered a typical quantitative analysis for biogenic amines, including histamine with the sensitive linear range from 2.5 to 100 ppm [24]. By modification, an HPLC method coupled with fluorescence detection was developed with linear over the range of 0.25–20 μg/mL for histamine and the detection limit of 75.0 μg/mL [7]. Kumar, N. [15] fabricated the silver nanoparticles capped with graphite as a sensor for determination of histamine and evaluated the linearity of 1–500 μM with the detection limit of 0.049 μM. Additionally, Leena Mattsson [24] applied a competitive, fluorescent, molecularly imprinted polymer (MIP) assay to detect histamine within the optimal linear range of 1–430 μM. Correspondingly, our Au-NPs probe provides a sensitive detection method of histamine via dual approaches. By the electrostatic reaction with histamine, the changes of electron density and molecular orbital induce the sensitive responses of Au-NPs, leading to the strong plasmon resonance absorption and amino-fluorophore vibrations/rotations. Moreover, Au-NPs offer a relatively broad range of 0.001–10.0 μM and 0.01–1.0 μM with a limit of detection of 0.87 nM and 2.04 nM by UV-vis and fluorescence spectrum assay, respectively. Moreover, the simplicity and rapidity of Au-NPs probes are two advantage factors in the detection process compared to the complex pretreatment process of traditional assays.

3.3. Selectivity of Au-NPs Probe

The chromogenic reaction and fluorescence response of Au-NPs was measured with various biological substances containing amino groups (histamine, putrescine, cadaverine, tyramine, phenylethylamine, guanine, guanosine, thymine, inosine, adenosine triphosphate (ATP), and adenosine monophosphate (AMP)) at concentrations of 1.0 μM each (Figure 4). Interestingly, the color of Au-NPs dramatically changed from wine red to dark blue by the target analytes of histamine, whereas other various biological substances containing amino groups still maintained their color (Figure 4A), likely due to histamine having a relatively high polar surface area, causing the bonding with Au-NPs. Meanwhile, the special UV-vis absorption spectra of Au-NPs bonded to histamine red-shifted from 520 nm to 664 nm, in sharp contrast with other biological substances, possibly due to the aggregation of Au-NPs induced by histamine (Figure 4C). By the electrostatic interactions and hydrogen-bonding between the negatively charged and positively charged histamine, the special UV-vis absorption peak of Au-NPs was red-shifted. Furthermore, only the solution of histamine-Au-NPs stood out in the experimental group with bright glaucous color under the UV-lamp (λ = 365 nm) (Figure 4B), and only the emission peak of that was significantly enhanced at 415 nm (Figure 4D), due to the fact that the electron density might have been increased by electrostatically attracted amino groups. Hence, histamine could be detected explicitly by Au-NPs via dual approaches of colorimetric and fluorometric perspectives.

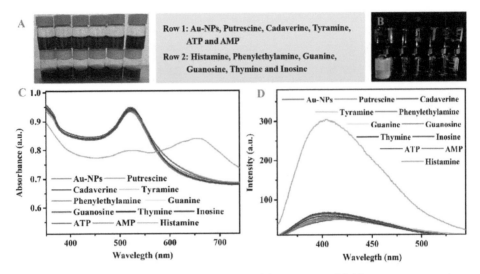

Figure 4. (**A**) colorimetric, (**B**) fluorescent response, (**C**) UV-vis spectra and (**D**)fluorescence spectra of Au-NPs added with different biological substances containing amino groups (histamine, putrescine, cadaverine, tyramine, phenylethylamine, guanine, guanosine, thymine, inosine, ATP, and AMP) at concentrations of 1.0 μM.

3.4. Detection of Histamine in Salmon Muscle

For further evaluation of the effectiveness of the Au-NPs probe in the real sample, the Au-NPs probe was experimentally applied to detect three various concentrations of histamine (0.01, 0.1, and 1.0 μM) in fish muscle (Table 1). Although there is still negligible interference in the actual sample analysis, the Au-NP method does not require complicated pretreatment and therefore has excellent stability. Au-NPs probe showed a satisfactory recovery value of 96.01%–103.43% with measurement precision (RSD < 10%) and accuracy (±4%), indicating Au-NPs probe can be competent with respect to the FDA and EU requirements for histamine testing in aquatic products. In addition, the application of Au-NPs probe analysis has simple preprocessing, suitable equipment requirements, and low experimental costs; thus, the efficient and convenient Au-NPs probe has potential application value in freshness and spoilage assay of aquatic food to ensure food safety.

Table 1. Analytical data for the colorimetric and fluorescence detection of histamine in real samples by Au-NPs probe, respectively.

Detection Methods		Spiked Concentration (μM)	Found Concentration (μM)	Recovery (%)	Precision (%)	Accuracy (%)
Au-NPs probe (this work)	Colorimetric detection	0.01	0.0102 ± 0.0010	102.13	9.32	2.13
		0.1	0.0997 ± 0.0049	99.67	4.95	−0.33
		1	1.0690 ± 0.0259	106.90	2.43	6.90
	Fluorescence detection	0.01	0.0096 ± 0.0004	96.01	4.46	−3.99
		0.1	0.1024 ± 0.0095	102.39	9.30	2.39
		1	1.0317 ± 0.0996	103.17	9.66	3.17

4. Conclusions

In summary, by adjusting the precursor ratio, an economical and practical Au-NP was fabricated to detect histamine via dual approaches of colorimetric and fluorescence perspectives with high sensitivity and selectivity. As histamine occurred, Au-NPs could precisely and quantitatively be detected by the response of UV-vis and fluorescence spectrums with a limit of detection of 0.87 nM and 2.04 nM, respectively. Moreover, the significant colorimetric and fluorescence turn-on phenomenon of Au-NPs involved with histamine could be captured by eye vision. Both of the two detection methods of Au-NPs were successfully applied to the quantitative analysis of histamine in fresh salmon muscles, suggesting that the simplicity and rapidity of the Au-NPs dual-detection method may be applicable to the FDA to judge the freshness and spoilage of aquatic foods.

Supplementary Materials: The following are available online at http://www.mdpi.com/2304-8158/9/3/316/s1, Scheme S1 Synthesis mechanism of Au-NPs, Figure S1 (A–D) TEM images of Au-NPs synthesized by different HAuCl$_4$/Na$_3$Ct precursor ratios (1:2, 1:4, 1:6, 1:8). (E) The average diameters and (F) UV-vis absorption spectra of Au-NPs. (G) Schematic illustration of the transformation process of Au-NPs size by citrate reduction.

Author Contributions: Conceptualization, J.B.; methodology, J.B. and G.-L.Z.; investigation, C.T.; data curation, H.H.; writing—original draft preparation and writing—review and editing, J.B.; supervision, H.-M.H.; project administration, H.-M.H.; funding acquisition, J.B. and H.-M.H. All authors have read and agreed to the published version of the manuscript.

Funding: This research was funded by National Basic Research Program of China (2019YFC1605904) and Natural Science Foundation of Liaoning Province (20180550157).

Acknowledgments: The authors gratefully acknowledge laboratory colleagues for their helps.

Conflicts of Interest: All authors declare no conflict of interest.

References

1. Heerthana, V.R.; Preetha, R. Biosensors: A potential tool for quality assurance and food safety pertaining to biogenic amines/volatile amines formation in aquaculture systems/products. *Rev. Aquacult.* **2019**, *11*, 220–233. [CrossRef]

2. Sørensen, K.M.; Aru, V.; Khakimov, B.; Aunskjær, U.; Engelsen, S.B. Biogenic Amines: A key freshness parameter of animal protein products in the coming circular economy. *Curr. Opin. Food Sci.* **2018**, *22*, 167–173. [CrossRef]

3. Del, R.B.; Redruello, B.; Linares, D.M.; Ladero, V.; Fernandez, M.; Martin, M.C.; Ruas-Madiedo, P.; Alvarez, M.A. The dietary biogenic amines tyramine and histamine show synergistic toxicity towards intestinal cells in culture. *Food Chem.* **2017**, *218*, 249–255. [CrossRef] [PubMed]

4. Yang, H.; Yoon, M.; Um, M.Y.; Lee, J.; Jung, J.; Lee, C.; Kim, Y.T.; Kwon, S.; Kim, B.; Cho, S. Sleep-Promoting Effects and Possible Mechanisms of Action Associated with a Standardized Rice Bran Supplement. *Nutrients* **2017**, *9*, 512. [CrossRef] [PubMed]

5. Altieri, I.; Semeraro, A.; Scalise, F.; Calderari, I.; Stacchini, P. European official control of food: Determination of histamine in fish products by a HPLC–UV-DAD method. *Food Chem.* **2016**, *211*, 694–699. [CrossRef] [PubMed]

6. Adams, F.; Nolte, F.; Colton, J.; De, B.J.; Weddig, L. Precooking as a Control for Histamine Formation during the Processing of Tuna: An Industrial Process Validation. *J. Food Protect.* **2018**, *81*, 444–455. [CrossRef]

7. Xinna, W.; Liang, Y.; Wang, Y.; Fan, M.; Sun, Y.; Liu, J.; Zhang, N. Simultaneous determination of ten kinds of biogenic amines in rat plasma using high-performance liquid chromatography coupled with fluorescence detection. *Biomed. Chromatogr.* **2018**, *32*, e4211.

8. Hui, Y.; Zhuang, D.; Hu, X.; Shuang, Z.; Zhiyong, H.E.; Zeng, M.; Fang, X.; Chen, J.; Chen, X. Rapid determination of histamine in fish by thin-layer chromatography-image analysis method using diazotized visualization reagent prepared with p-nitroaniline. *Anal. Methods-UK* **2018**, *10*, 3386–3392.

9. Kamankesh, M.; Mohammadi, A.; Mollahosseini, A.; Seidi, S. Application of a novel electromembrane extraction and microextraction method followed by gas chromatography-mass spectrometry to determine biogenic amines in canned fish. *Anal. Methods-UK* **2019**, *11*, 1898–1907. [CrossRef]

10. Selnihhin, D.; Sparvath, S.M.; Preus, S.; Birkedal, V.; Andersen, E.S. Multi-Fluorophore DNA Origami Beacon as a Biosensing Platform. *ACS Nano* **2018**, *12*, 5699–5708. [CrossRef]

11. Han, L.; Liao, J.D.; Sivashanmugan, K.; Liu, B.; Fu, W.; Chen, C.C.; Chen, G.D.; Juang, Y.D. Gold Nanoparticle-Coated ZrO2-Nanofiber Surface as a SERS-Active Substrate for Trace Detection of Pesticide Residue. *Nanomaterials* **2018**, *8*, 402–413.

12. Rong, Y.; Shou, Z.; Chen, J.; Wu, H.; Zhao, Y.; Qiu, L.; Jiang, P.; Mou, X.Z.; Wang, J.; Li, Y.Q. On–Off–On Gold Nanocluster-Based Fluorescent Probe for Rapid Escherichia coli Differentiation, Detection and Bactericide Screening. *ACS Sustain. Chem. Eng.* **2018**, *6*, 4504–4509.

13. Rkr, G.; Palathedath, S.K. Cu@Pd core-shell nanostructures for highly sensitive and selective amperometric analysis of histamine. *Biosens. Bioelectron.* **2018**, *102*, 242–246.

14. Toloza, C.A.T.; Khan, S.; Silva, R.L.D.; Romani, E.C.; Larrude, D.G.; Louro, S.R.W.; Júnior, F.L.F.; Aucélio, R.Q. Photoluminescence suppression effect caused by histamine on amino-functionalized graphene quantum dots with the mediation of Fe^{3+}, Cu^{2+}, Eu^{3+}: Application in the analysis of spoiled tuna fish. *Microchem. J.* **2017**, *133*, 448–459. [CrossRef]

15. Kumar, N.; Goyal, R.N. Silver nanoparticles decorated graphene nanoribbon modified pyrolytic graphite sensor for determination of Histamine. *Sens. Actuators B-Chem.* **2018**, *268*, 383–391. [CrossRef]

16. Kuo, P.C.; Lien, C.W.; Mao, J.Y.; Unnikrishnan, B.; Chang, H.T.; Lin, H.J.; Huang, C.C. Detection of urinary spermine by using silver-gold/silver chloride nanozymes. *Anal. Chim. Acta* **2018**, *1009*, 89–97. [CrossRef]

17. Beibei, L.; Hang, G.; Wang, Y.; Zhang, X.; Pan, L.; Qiu, Y.; Wang, L.; Hua, X.; Guo, Y.; Wang, M.; et al. A gold immunochromatographic assay for simultaneous detection of parathion and triazophos in agricultural products. *Anal. Methods-UK* **2018**, *10*, 422–428.

18. Shehata, D.M.; Hadi, S. Applications of gold nanoparticles in virus detection. *Theranostics* **2018**, *8*, 1985–2017.

19. Baetsen-Young, A.M.; Vasher, M.; Matta, L.L.; Colgan, P.; Alocilja, E.C.; Day, B. Direct colorimetric detection of unamplified pathogen DNA by dextrin-capped gold nanoparticles. *Biosens. Bioelectron.* **2018**, *101*, 29–36. [CrossRef]

20. Yan, L.; Ding, D.; Zhen, Y.; Guo, R. Amino acid-mediated 'turn-off/turn-on' nanozyme activity of gold nanoclusters for sensitive and selective detection of copper ions and histidine. *Biosens. Bioelectron.* **2017**, *92*, 140–146.

21. Ping, S.; Wang, G.; Kang, B.; Guo, W.; Liang, S. High-Efficiency and High-Color-Rendering-Index Semitransparent Polymer Solar Cells Induced by Photonic Crystals and Surface Plasmon Resonance. *ACS Appl. Mater. Interfaces* **2018**, *10*, 6513–6520.

22. Dong, Y.; Pang, H.; Yang, H.B.; Guo, C.; Shao, J.; Chi, Y.; Li, C.M.; Yu, T. Carbon-based dots co-doped with nitrogen and sulfur for high quantum yield and excitation-independent emission. *Angew. Chem. Int. Ed. Engl.* **2013**, *52*, 7800–7804. [CrossRef] [PubMed]

23. Fujiki, A.; Uemura, T.; Zettsu, N.; Akai-Kasaya, M.; Saito, A.; Kuwahara, Y.J.A.P.L. Enhanced fluorescence by surface plasmon coupling of Au nanoparticles in an organic electroluminescence diode. *Appl. Phys. Lett.* **2010**, *96*, 14. [CrossRef]

24. Mattsson, L.; Xu, J.; Preininger, C.; Bui, B.T.S.; Haupt, K. Competitive fluorescent pseudo-immunoassay exploiting molecularly imprinted polymers for the detection of biogenic amines in fish matrix. *Talanta* **2018**, *181*, 190–196. [CrossRef]

Article

FT-NIR Analysis of Intact Table Grape Berries to Understand Consumer Preference Driving Factors

Teodora Basile *, Antonio Domenico Marsico, Maria Francesca Cardone, Donato Antonacci and Rocco Perniola

Consiglio per la ricerca in agricoltura e l'analisi dell'economia agraria-Centro di ricerca Viticoltura ed Enologia (CREA-VE), Via Casamassima 148-70010 Turi (Ba), Italy; adomenico.marsico@crea.gov.it (A.D.M.); mariafrancesca.cardone@crea.gov.it (M.F.C.); donato.antonacci@crea.gov.it (D.A.); rocco.perniola@crea.gov.it (R.P.)
* Correspondence: teodora.basile@crea.gov.it; Tel.: +39-080-8915711; Fax: +39-080-4512925

Received: 5 December 2019; Accepted: 15 January 2020; Published: 17 January 2020

Abstract: Fourier-transform near infrared spectroscopy (FT-NIR) is a technique used in the compositional and sensory analysis of foodstuffs. In this work, we have measured the main maturity parameters for grape (sugars and acids) using hundreds of intact berry samples to build models for the prediction of these parameters from berries of two very different varieties: "Victoria" and "Autumn Royal". Together with the chemical composition in terms of sugar and acidic content, we have carried out a sensory analysis on single berries. Employing the models built for sugars and acids it was possible to learn the sweetness and acidity of each berry before the destructive sensory analysis. The direct correlation of sensory data with FT-NIR spectra is difficult; therefore, spectral data were exported from the spectrometer built-in software and analyzed with R software using a statistical analysis technique (Spearman correlation) which allowed the correlation of berry appreciation data with specific wavelengths that were then related to sugar and acidic content. In this article, we show how it is possible to carry out the analysis of single berries to obtain data on chemical composition parameters and consumer appreciation with a fast, simple, and non-destructive technique with a clear advantage for producers and consumers.

Keywords: NIR; PLS; PCA; correlogram; sensory analysis

---

1. Introduction

The choice of proper harvest time is especially important in non-climacteric fruits such as grape. The maturity requirements for table grape commercialization rely only on two main parameters: sugars (soluble solid concentration, ° Brix) and acid content (titratable acidity), or their ratio (° Brix/acid ratio) [1–3]. Even if the content of sugars and acids is a valuable predictor of quality, there are other factors that strongly influence sensorial judgment [4]. Indeed, the evaluation of table grapes in terms of consumer acceptability is better performed through a sensorial analysis [5]. Among non-destructive techniques, FT-NIR spectroscopy is one of the most advanced concerning instrumentation, applications, accessories, and chemometric software packages [6]. FT-NIR spectroscopy combined with chemometrics has been used as a tool for geographical origin discrimination [7] and quality control [8,9] since this technique allows the development of calibration models that can be used for both the quantitative and qualitative determination of various parameters (e.g., pH, sugars, etc.), including minor components (e.g., polyphenols) [10,11] in various food matrices and beverages. Since FT-NIR spectroscopy can capture variations of both chemical and sensory properties, this technique seems ideal to investigate the influence of sugars and acidic composition on consumer acceptance. As reported in previous studies on intact grape berries, no direct correlation has been found between NIR spectra and sensory analysis [12]. Moreover, the majority of articles involving FT-NIR analysis of grape focuses on berry juice, homogenates, or skin extracts. Scanning of single berries is also possible; however, high coefficients of variation in the FT-NIR spectra are observed when samples are scanned in different positions relative to the FT-NIR source. This variation within the same berry might be linked to dust, different degrees of sun exposure, or simply to variations in the punctual chemical composition in terms of sugar and acidic content. Therefore, a better prediction for a homogenized sample compared to intact berry is usually obtained. To overcome the issues linked to inhomogeneity a larger number of samples must be used to build the NIR prediction models for each variety analyzed. In this work, intact berries of two very different table grape varieties (a white-seeded one and a red seedless one), namely "Victoria" and "Autumn Royal", have been analyzed using different chemical analytical techniques in conjunction with sensory analysis. Chemometric techniques have been applied to search for a correlation between chemical composition in terms of sugar and acidic content and sensory data. A correlogram graph using Spearman coefficients was built and employed to search for a relationship (in terms of higher correlation) between sensory data and wavenumbers in the NIR spectrum. The attribution of those wavelengths to specific compounds (sugars or acids) was tentatively performed. When a correlation was not found between acceptance and the wavelengths linked to the maturity parameters investigated, we hypothesized that other features such as the texture and appearance of grape, which are known to have an important influence in the assessment of fruit, could have been responsible for the positive acceptance.

2. Material and Methods

2.1. Chemical Analysis

The table grape berries were collected during the 2018 vintage from the experimental vineyards of CREA Research Centre for Viticulture and Enology of Turi, Southern Italy and were obtained from local companies leading producers of table grape. Two very different table grape varieties: a white-seeded one and a red seedless one, namely "Victoria" and "Autumn Royal", with an range of maturity levels below and above the minimum required values for grape commercialization (12.5° Brix given a total soluble solid content (TSS)/titratable acidity (TA) over 20:1) [1–3] were analyzed. After NIR spectra acquisition, the main parameters for grape quality evaluation were analyzed with the primary methods described in the following lines. Total soluble solid content (TSS, ° Brix) was determined at 20 °C using a digital refractometer Atago PR1 (Atago Co., Tokyo, Japan). Titratable acidity (TA) was measured as tartaric acid (g/L) by titration of grape juice with sodium hydroxide (0.1 N) to an endpoint pH of 7. An HPLC analysis of organic acids was performed on single berry juice using an HP 1100 apparatus (Agilent, Palo Alto, CA, USA) with a diode array detector (DAD) set at 210 nm and a Synergi Hydro-RP 80A column, 250 × 4.6 mm, 4 µm (Phenomenex Inc., Torrance, CA, USA) as stationary phase [13]. A semi-automated method for colorimetric and enzymatic assays was employed for sugar determination on berry juice (SATURNO 150 Crony Instruments). The analysis kit employed was specific for D-fructose and D-glucose. The amount of NADPH formed through the combined action of hexokinase (HK), phosphoglucose isomerase (PGI), and glucose-6-P dehydrogenase (G6PDH), measured at 340 nm, is stoichiometric with the amount of D-glucose and D-fructose in the sample.

2.2. NIR Analysis

A Bruker TANGO FT-FT-NIR spectrometer was employed for spectra acquisition and OPUS/QUANT software (Bruker Optik GmbH, Ettlingen Germany) Vers. 2.0 was used for chemometric analysis. In order to build FT-NIR calibration models for the investigated parameters, after the NIR analysis hundreds of samples of fresh table grape for each variety were analyzed with primary methods. Each berry was rinsed in distilled water and gently wiped with paper prior to NIR analysis in order to remove any surface dust and dirt without compromise the superficial wax coating of berries. Each berry was placed on the flat surface of a sample cup with a quartz window and manually rotated in order to record the FT-NIR spectra on three different berry faces. Due to the small dimensions and round shape of the samples, it was not possible to automatically rotate the sample; therefore after each measurement, the berry was manually moved. Since berries were not always able to cover the whole emission source inlet of the instrument, to avoid the record of signals from the air each sample was covered with a hollow metal tube. A background spectrum was automatically recorded prior to each sample while both temperature and humidity were kept constant. After the selection of the most relevant wavelengths to reduce the prediction error associated with spectra noise, different pre-processing techniques were compared in order to eliminate unnecessary physical information and magnify relevant variations in the original spectra. The first derivative (FD) and the vector normalization (VN) were chosen in accordance with their predictive performance for the investigated parameters. With the OPUS/QUANT Bruker software the partial least squares (PLS) regression approach was used for the quantification of changes in the sugar, acid, or sensory-related parameters. After cross-validation, outlier removal, and optimization steps, the final version of the calibration models was obtained with the same Bruker software.

2.3. Sensory Characterisation

In order to evaluate preferences among grapes with different degrees of ripeness, a hedonic test was performed [14]. Our panel test was composed of 82 subjects (58 high school students and 24 adults) recruited from subjects already involved in a project with the Council for Agricultural Research and Analysis of Agricultural Economics (CREA) Research Centre for Viticulture and Enology and from the staff of the same research center. Berries for tasting were taken from the central part of each bunch, washed in order to remove dirt and dust, shortly wiped on paper, and kept at 20 °C. The room temperature was set at 20 °C and the tasting was performed in the morning (from 10:00 to 12:00 h). After FT-NIR analysis of single berries, each berry was placed on an identical white plate coded with a two-digit random number and presented in random order. Tasters were asked to evaluate the likeability of grapes with different maturity levels without communicating with each other. Filtered water and water crackers were provided as palate cleansers. Each taster was given a set of each variety of grape samples at different ripeness level, without information about differences, and was asked to rate the samples on a scale of 1 to 10. The significance of the sensory analysis was assessed performing the Kruskal–Wallis non-parametric test followed by the Dunn test as a post hoc test, using a p-value < 0.05 [14]. Moreover, in order to highlight the taster's preference, consumers' acceptance (%CA) was calculated as follows: number of tasters giving a preference value over 5 divided by the total number of tasters [4].

2.4. Structure of Data Sets and Statistical Analysis

In order to search for a relationship between FT-NIR and sensory data, a data set was built. Two FT-NIR data matrices (N_{1i}), one for each variety analyzed ($i = 2$), were dimensioned at 350×1900. Rows represented the number of samples analyzed (50 berries for each of the seven maturity levels) and columns the parameters for FT-NIR measurements (1899 absorbance values recorded with a 4 cm^{-1} step in the 11,544–3952 cm^{-1} range and one descriptor). N_{2i} represents the corresponding mean values matrices dimensioned at 7×1900. Sensory data were collected in two matrices: An S_{1i} matrix was dimensioned at 1722×1 in which the rows represented the mean score of preference for each of the maturity levels (seven maturity levels, three berries for each maturity level, 82 judges) and the column was the preference parameter. Moreover, an S_{2i} matrix dimensioned at 7×1 (mean values for each of the seven maturity levels × one descriptor) was created. Principal component analyses (PCAs) on the N_{1i} matrix as it was and after the removal of wavelengths with negative correlation with "preference" sensory parameter were performed for the "Victoria" variety. PCAs were performed on the average data matrices (N_{2i}) for both "Victoria" and "Autumn Royal" varieties. Moreover, a correlation analysis was performed on the average data matrix obtained by the concatenation of N_{2i} and S_{2i} for each variety. All the statistical procedures described in detail were carried out using the R software environment [15].

3. Results and Discussion

3.1. Chemical Analysis

Any direct attribution of specific vibrations to the molecules of sugars was hindered by the nature of the sample. It was not possible to perform an NIR analysis of laboratory-prepared samples containing known concentrations of the compound of interest (even in a sample-like medium) since we were measuring intact berries. Moreover, it was not possible to use an internal standard to enhance and thus make more easily identifiable peaks related to a specific overtone vibration to the intact berry. We could have added a sugar (e.g., glucose with a selective deuterium exchange for sugar OH groups) to the juice and then measure the spectrum of the modified sample, but fruit juice spectra and those of intact berry differ. Therefore, as was done by other authors in previous papers, a tentative peak assignment of FT-NIR spectra was done in accordance with the literature. Prominent FT-NIR absorption peaks were observed around 10,200, 8400, 6900, and 5600 cm^{-1} (Figure 1).

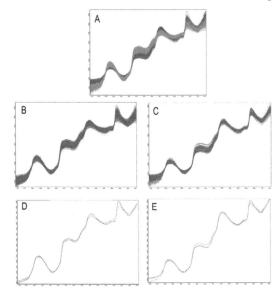

Figure 1. (**A**) Superimposed normalized Fourier-transform near infrared spectroscopy (FT-NIR) spectra (absorbance vs. wave numbers) of single berries for "Autumn Royal" (blue) and "Victoria" (red). Normalized spectra with maturity level-based coloration: "Autumn Royal" (**B**) and "Victoria" (**C**). Averaged spectra for each maturity level: "Autumn Royal" (**D**) and "Victoria" (**E**).

Whereas many compounds absorb in the FT-NIR bands highlighted here, those at 10,200 and 6900 cm^{-1} are related to water (second overtone and first overtone of O-H of water, respectively) [16]; this is usually the case for fruits and vegetables and their juices with 70%–80% of water [17]. Sugars and organic acids show absorptions in the same regions of 6978, 5643, and 5614 cm^{-1} [18]. The absorption bands at 6900 cm^{-1} are also related to a combination of stretch and deformation of the O-H group in glucose [19,20]. The 5643 and 5614 cm^{-1} wavelengths are related to the first overtone of CH3 methyl, the first overtone of CH2 methylene, and the first overtone of CH aliphatic, and are attributed to vibrations of the molecules of sugars [21]. C-H overtone and stretch in sugars and organic acids were observed at 8436 and 6978 cm^{-1}, whilst the C-O stretch and overtone in sugars and organic acids were attributed to regions at 6978, 5643, and 5614 cm^{-1} [22]. FT-NIR calibration models were built for the investigated chemical parameters. The output of OPUS/QUANT software analysis shows the calibration model together with several statistical parameters such as the root mean square error of cross validation (RMSECV), coefficient of determination (R^2), and bias. These are a measure for the deviation of the predicted values from the actual values obtained by reference analytical methods. These prediction models could be employed to predict sugar and acid composition of intact berries. The spectra were recorded in the 12,000–3600 cm^{-1} range; a background spectrum was automatically recorded prior to each sample while both temperature and humidity were kept constant. Different pre-processing techniques were compared in order to eliminate unnecessary physical information and magnify relevant variations in the original spectra. The first derivative (FD) together with the vector normalization (VN) was chosen in accordance with their predictive performance. The prediction model obtained with the OPUS/QUANT software (Bruker Optik GmbH, Ettlingen Germany) Vers. 2.0. for TSS shows good predictability (R^2 = 83%) (Table 1).

Even if the same number of samples was employed for both TSS and TA FT-NIR analysis, the number of spectra employed to build the model for TA was considerably smaller than those employed for TA analysis. This resulted from the removal of several spectra from the calibration data set which were considered outliers. For TA the predictability of the resulting model was not as good as for TSS. We hypothesize that this lack of prediction ability of FT-NIR can be linked to the gradient of acidity normally present in grape berries (it increases towards the center). Probably, the FT-NIR analysis does not always reach the layer which contains the main acidic compounds; therefore, the acidity value is not accurate enough for the sensitivity of the FT-NIR analysis. We performed the analysis of single organic acids and sugars with HPLC and enzymatic assay. The prediction models built with FT-NIR data did not give satisfactory results for both individual organic acids and sugars. Probably the reason for this lack of predictivity for organic acids is the same as for TA values. Concerning sugars, the number of samples employed was not adequate to determinate a suitable FT-NIR response or the analytical method employed gives a value which is not accurate enough for the sensitivity of the FT-NIR analysis. Under this second hypothesis, a change in the primary method employed for sugar determination would result in a better prediction for single berries. Concerning the "preference" sensory parameter it was not possible to find a good correlation with the PLS regression; therefore, it was not possible to build a suitable model to predict this sensory parameter from NIR spectral features (Figure 2).

Table 1. Quality of the models obtained with the PLS algorithm. TSS: total soluble solid content; TA: titratable acidity; RMSECV: root mean square error of cross validation.

Parameter	RMSECV	Rank	Regions (cm^{-1})	Preprocessing	R^2 (%)	Bias
TA	0.861	5	9400–6100 5452–4600	Vector normalization	57.32	−0.000762
TSS	1.3	6	9400–7500	First derivative + Vector normalization	83.04	0.01

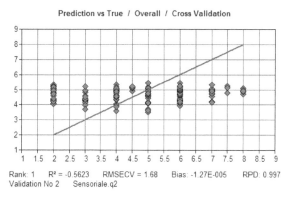

Prediction vs True / Overall / Cross Validation

Rank: 1 R^2 = -0.5623   RMSECV = 1.68   Bias: -1.27E-005   RPD: 0.997
Validation No 2   Sensoriale.q2

Figure 2. Output of OPUS/QUANT software: Validation curve of "preference" parameter from sensory analysis.

3.2. Sensory Data Analysis

The analysis of single grape berries with analytical techniques or sensory analysis was performed in conjunction with the record of FT-NIR spectra. This procedure allowed us to have information concerning the chemical composition in terms of sugar and acidic content of each berry (from NIR prediction models) that was employed in the destructive sensory analysis. Sensory data collected from the hedonistic test were analyzed for panel acceptance following the equation: Percentage of consumer acceptance = Number of panelists with a rating > 5/Total number of panelists [4]. From Table 2 it is possible to notice that the consumer acceptance does not decrease linearly with the decrease of ° Brix (nor with the TSS/TA ratio). It is particularly interesting that for the different varieties ("Autumn Royal" and "Victoria" grapes) the acceptance was quite high even with a low sugar content.

Table 2. Maturity parameters and consumers' acceptance.

Variety	Maturity Parameters [†]			Sensory Data [‡]	
	° Brix	TA	TSS/TA	%Acceptance	Preference
	18.3 ± 0.1 [a]	3.8 ± 0.1 [e]	49	100	7.13 ± 0.63 [a]
	16.4 ± 0.1 [b]	4.5 ± 0.1 [d]	36	81	6.42 ± 0.58 [ab]
	14.4 ± 0.0 [c]	5.5 ± 0.1 [ab]	26	50	5.63 ± 0.85 [ab]
Autumn royal	13.8 ± 0.1 [d]	5.0 ± 0.1 [c]	27	56	5.17 ± 0.68 [ab]
	13.3 ± 0.1 [e]	5.0 ± 0.1 [c]	27	33	5.25 ± 1.41 [ab]
	12.6 ± 0.2 [f]	5.7 ± 0.1 [a]	22	29	4.40 ± 1.29 [b]
	11.8 ± 0.2 [g]	5.4 ± 0.1 [b]	22	44	4.56 ± 1.88 [b]
	° Brix	TA	TSS/TA	%Acceptance	Preference
	16.9 ± 0.1 [a]	3.8 ± 0.1 [d]	45	86	6.72 ± 1.85 [a]
	14.9 ± 0.1 [b]	3.9 ± 0.1 [d]	38	86	6.08 ± 1.23 [a]
	14.2 ± 0.1 [b]	5.4 ± 0.1 [c]	26	87	5.80 ± 1.50 [a]
Victoria	11.9 ± 0.1 [c]	5.3 ± 0.0 [c]	22	40	4.20 ± 0.60 [b]
	11.6 ± 0.1 [c]	3.0 ± 0.0 [e]	39	65	4.70 ± 0.50 [b]
	10.6 ± 0.0 [d]	6.7 ± 0.2 [b]	16	32	4.08 ± 1.81 [b]
	9.0 ± 0.1 [e]	10.7 ± 0.1 [a]	8	4	2.41 ± 1.29 [b]

Values are mean ±SD. For each variety values in the same column bearing different letters are significantly different. Data were analyzed with [†] an ANOVA test followed by a Tukey post hoc test or [‡] the Kruskal–Wallis non-parametric test followed by the Dunn post hoc test using a p-value < 0.05.

Indeed, palatability is a complex result of not only main (sugars and acids) and minor (polyphenols, tannins) components but also other "not flavoring" properties like color and texture related sensations e.g., crunchiness, gumminess, etc. [23]. Among other factors, it is known how the visual appearance of berries is the first quality index for consumers. Indeed, varieties such "Victoria" and "Autumn Royal" with well-defined skin pigmentation even at the beginning of maturation (a grape variety characteristic) show high acceptance values at low glucidic content. Other authors suggest that grape texture is another very important factor in quality judgment and could be used as an indicator for ripening. Indeed, on other fruits such as apples or tomatoes, texture is the main characteristic for determining fruit maturity [23]. In order to understand the importance of sugar and acidic composition and to investigate the presence of other factors influencing the panel judgment on the preference of consumers a deeper statistical analysis was performed (Spearman non-parametric test) on sensory and analytical data together (see Paragraph 3.3).

3.3. Chemometrics

3.3.1. PCA

The PCA analysis is a statistical discrimination technique able to highlight common features that allow to group samples with similar composition and thus underlying attributes as shown in previous works [24,25]. Since the PCA performed on the spectra with the spectrometer built-in software (OPUS, Bruker) did not highlight specific features (data not shown), a novel MACRO was created on the OPUS software which allowed for the export of hundreds of spectra files at once in an R readable format. These spectra were processed (truncated and normalized i.e., vector normalization and first derivative) in the same way as those employed in the PLS procedure. From each spectrum, the absorbance values for each of the wave numbers were then inserted in a matrix ($n \times 1900$; number of samples × wave number values) and this matrix was employed for further statistical analyses. Since it was not possible to easily differentiate groups of samples with different maturity levels we performed a PCA on averaged spectra for each maturity level. Averaging can reduce noise without compromising data while effectively preserving main features which are responsible for differentiation among samples with different maturity levels. The PCA performed on the mean values of absorbance using the whole FT-NIR spectral window for "Victoria" grape samples with different maturity levels is shown in Figure 3.

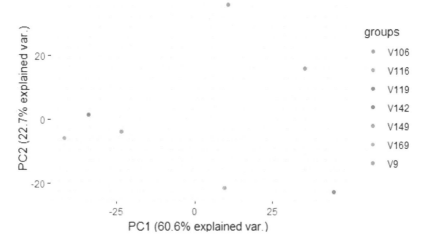

Figure 3. Principal component analysis (PCA) of mean values for the 1900 absorbance values of FT-NIR spectra for the seven maturity levels of "Victoria" grape: 16.9° (V106), 14.9° (V149), 14.2° (V142), 11.9° (V119), 11.6° (V116), 10.6° (V106), and 9° (V9) Brix.

Grouping found in the PCA reflects the maturity levels of the samples. Indeed, the lowest maturity samples are grouped in the positive quadrant of the PCA, the intermediate (around 11° Brix) in the second quadrant, the two highest maturity samples are in the negative quadrant, and the 14.2° Brix sample is in the fourth quadrant. Large spectral regions give a strong contribution to the positive PC1 axis: 11,548–10532, 9600–9112, 7208–6488, 5328–4772, and 4228–3952 cm^{-1}. Three of these regions can be attributed to water signals; indeed the 10,500 and 6900 cm-1 regions are related to water [16,17], and the 5000–5300 cm^{-1} region is also attributable to water [26]. It looks like that the water content mainly contributes to the positive *x*-axis of the PCA plot. The interpretation of the main contributors of the PC2 axis is quite hard since the loadings are quite low. Among the spectral regions which give a higher contribution to the negative PC2 axis, there is one (4988–4308 cm^{-1}) that is likely related to carbohydrates and organic acids [26–29]. Therefore, we can assume that samples with a negative *y*-axis are characterized by a higher contribution of carbohydrates and organic acids. For the "Autumn Royal" variety a PCA was performed with the mean values of absorbance using the whole FT-NIR spectral window. The PCA (Figure 4) shows that "Autumn Royal" grape samples with 13.8° and 14° Brix are closely related (positive quadrant) as well as the 11° and 12° Brix ones (fourth quadrant), which was expected since the maturity parameters do not differ as much as for the other maturity levels.

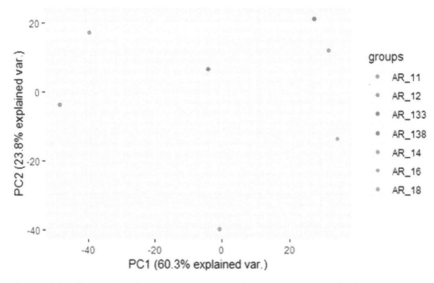

Figure 4. PCA of mean values for the 1900 absorbance values of FT-NIR spectra for the seven maturity levels of "Autumn Royal" grape: 11° (AR11), 12° (AR12), 13.3° (AR133), 13.8° (AR138), 14° (AR14), 16° (AR16), and 18° (AR18) Brix.

The other maturity levels are clearly differentiated, with the 18° Brix sample completely separated from the others, with a highly negative value of the PC2 axis, and one close to zero for the PC1 axis. Among the main wave numbers contributing to the negative PC2 (ranging from −0.5 to −0.9), there are regions at 6772–5324 and 4706–4256 cm^{-1} which can both be attributed to sugar signals [26,30]. Therefore, concerning the PC2 axis, it looks like the influence of sugar content was responsible for the sample's differentiation. The main FT-NIR regions contributing to the negative PC1 axis (even if with a weight around −0.2) include three of the prominent FT-NIR absorption peaks: 10,200, 8436, and 5614 cm^{-1} (see Figure 1). The absorption at 10,200 is related to water [16], the other two regions are related to sugars and organic acids which show absorptions in the same regions [16,22]. A higher contribution of organic acids and water to the negative PC1 axis explains the grouping of lower maturity samples on the left of the PCA plot.

3.3.2. Correlogram

The prediction models built using FT-NIR spectroscopy data in conjunction with sensory analyses did not give satisfactory results, as was previously found in the literature for intact grape samples [12,23]. Even if a direct prediction of sensory attributes from instrumental data was not possible, the exported FT-NIR spectral data were employed to perform deeper statistical analyses. Even a slight difference in grape composition is captured by FT-NIR spectra; therefore, even grape berries of the same variety belonging to the same "maturity level" show different spectral characteristics. To find common features among grape with similar TSS and TA values, which are the only parameters that must satisfy a specific minimum value for grape commercialization, the hundreds of spectra recorded for each maturity level of each variety were averaged. Therefore, the search for a correlation between sensory and instrumental parameters for both "Autumn Royal" and "Victoria" varieties was performed using averaged NIR spectra of samples with the same acidity and sugar content and corresponding acceptance values. To analyze the correlation for each variety a matrix was created binding each N_{2i} matrix (mean values) to a vector containing the mean preference values for each maturity level of the specific variety. To search for a statistically significant relationship (i.e., to measure the existence and the strength of the relationship) between absorbance at each wavelength and acceptance values, the appropriate method is the Spearman rank correlation, since our variables are not normally distributed. To make the outputs more readable, the string containing the Spearman coefficients for sensory descriptor vs. wave-number (the element m_{xj} of the Spearman correlation matrix M_i representing the correlation coefficient of the sensory descriptor x and the wavelength j) was plotted in a "correlogram" graph (Figure 5).

Victoria Preference

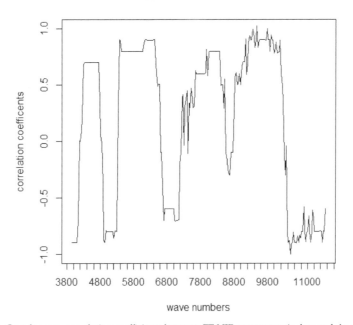

Figure 5. Correlogram: correlation coefficients between FT-NIR spectroscopic data and the sensory data "preference" of "Victoria" grape.

The highest correlation in the graph of the correlation coefficients between preference and each wavelength of the FT-NIR spectra in Figure 5 was associated with wave numbers in the following regions of the FT-NIR spectra: 4000–4500, 5500–6500, and 9000–10,000 cm^{-1}, which are all likely related to overtones of carbohydrates and organic acids [16,17,26,28,29]. Therefore, it seems that the sugar and acidic composition of tested grapes directly influenced the preference of the panel. The graph of the correlation coefficients between preference and each wavelength of the FT-NIR spectra for "Autumn Royal" (Figure 6) showed the highest correlation coefficient ($R = -0.97$) for 5372 cm^{-1}; this is likely linked to sugars since spectral regions between 5800 and 5400 cm^{-1} are related to the first overtone of C-H stretching and C-H + C-H and C-H + C-C combination bands, respectively, both attributed to vibrations of the molecules of sugars [26–29,31].

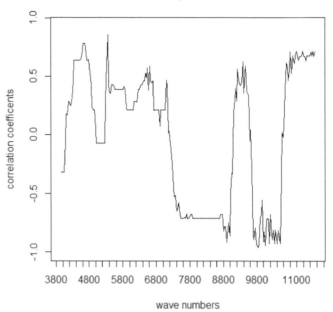

Figure 6. Correlogram: correlation coefficients between FT-NIR spectroscopic data and the sensory data "preference" of "Autumn Royal" grape.

Moreover, the region from 4600 to 4000 cm^{-1} can be ascribed to combinations of O-H bend/hydrogen-bonded O-H stretch (4428 cm^{-1}), O-H stretch/C-C stretch (4393 cm^{-1}), and combinations of C-H/C-C (4385–4063 cm^{-1}) vibrations of the sugar molecules [21]. Besides the positive correlation for the low wave numbers regions, the rest of the correlogram of "Autumn Royal" differs strongly from the "Victoria" one. Strong peaks between 7400 and 6400 cm^{-1} are mainly related to the first overtone of O−H stretching bands of water and while for "Victoria" they are highly negatively correlated (−0.7), for "Autumn Royal" just a slightly low positive correlation (+0.25) is observed. The other regions showing different correlations are: 8500–7500 cm^{-1} positive for "Victoria" and negative for "Autumn Royal", 9610–10,300 cm^{-1} positive for "Victoria" and negative for "Autumn Royal", and 10,490–11,548 cm^{-1} negative for "Victoria" and positive for "Autumn Royal". It seems clear that the preference for "Victoria" and "Autumn Royal" is not correlated to the same spectral signals and thus it was not influenced by the same grape features. This difference is not surprising because these two cultivars are very different in terms of overall appearance (one is red and seedless while the other is white seeded) and composition.

4. Conclusions

The objective of this research was the analysis of intact table grape berries to search for a relationship between the chemical composition in terms of sugar and acidic content and the consumers' preference using near-infrared spectroscopy. The analysis of single berries with analytical techniques or sensory analysis was performed in conjunction with the record of FT-NIR spectra; this allowed to us obtain information on the chemical composition in terms of sugar and acidic content of each berry and, at the same time, to link it to the sensory appreciation of the berry. To search for a correlation between consumers' preference parameter and chemical composition in terms of sugar and acidic content a correlogram graph was built for each of the two different grape varieties analyzed: "Victoria" (white-seeded) and "Autumn Royal" (red seedless). Vibrational transitions of different chemical compounds in the sample are responsible for absorbance peaks at different wavenumbers of a NIR spectrum. Since in the correlograms of the two varieties the spectral regions with a higher correlation with preference were different, it seems that the reason behind consumers' appreciation of these two different varieties can be related to different grape characteristics. The attribution of NIR spectra to specific compounds is not simple due to overlapping signals of different groups and the presence of strong absorbing molecules (e.g., water) in fruit samples. A PCA analysis performed for each variety using the absorbance values of the FT-NIR spectra as factors showed how it was possible to group grape of the same variety based on different values of TSS and TA. We searched for a possible explanation of the grouping observed in the PCA plot based on the wave-number regions with higher loadings on the PCA plot axes which have been linked to water, sugar, and acidic composition. Therefore, the PCA analysis was then employed as a tool to link wavelength regions to compounds responsible for the signals in the NIR spectra. The wavelength regions and chemical compounds responsible for those signals were used to interpret the correlograms. It was observed that while for the correlogram of "Victoria" variety sugars and acid content did influence the appreciation, for "Autumn Royal" the appreciation was not strongly correlated to wavelengths linked to sugar and acids; therefore, for this latter variety, the appreciation mainly relies on other parameters. The results of this work show how it is possible to extract valuable information from NIR spectra and search for a link between chemical composition in terms of sugar and acidic content and quality parameters even from intact berries using adequate chemometric techniques. This paper provides a good basis for more elaborate studies to correlate sensory properties and chemical composition of food.

Author Contributions: Conceptualization: T.B. and A.D.M.; Data curation: A.D.M.; Formal analysis: T.B.; Methodology: R.P.; Resources: M.F.C. and D.A.; Software: A.D.M.; Supervision: D.A.; Validation: R.P.; Writing—original draft, T.B.; Writing—review and editing: A.D.M., M.F.C., and R.P. All authors have read and agreed to the published version of the manuscript.

Funding: This research was funded by International Organisation of Vine and Wine (OIV) grant number [OIV 2017].

Acknowledgments: This work was supported by the International Organization of Vine and Wine (OIV) 2017 research grant.

Conflicts of Interest: The authors declare no conflict of interest.

References

1. OIV (International Organisation of Vine and Wine). Resolution VITI 1/2008 Standard on Minimum Maturity Requirements for Table Grapes. Available online: http://www.oiv.int/public/medias/369/viti-2008-1-en.pdf (accessed on 1 February 2019).
2. Food and Agriculture Organization of the United Nations (FAO). *Codex Standard for Table Grapes (CODEX STAN 255-2007) Amended 2011.* Available online: http://www.fao.org/fao-who-codexalimentarius/sh-proxy/en/?lnk=1&url=https%253A%252F%252Fworkspace.fao.org%252Fsites%252Fcodex%252FStandards%252FCXS%2B255-2007%252FCXS_255e.pdf (accessed on 1 February 2019).
3. European Union (EU). Commission Implementing Regulation (EU) No 543/2011 of 7 June 2011.
4. Jayasena, V.; Cameron, I. Brix/acid ratio as a predictor of consumer acceptability of Crimson seedless table grapes. *J. Food Qual.* **2008**, *31*, 736–750. [CrossRef]
5. OIV (International Organisation of Vine and Wine). Resolution OIV/VITI 371/2010 OIV General form for the Sensorial Analysis of Table Grapes. Available online: http://www.oiv.int/public/medias/385/viti-2010-2-en.pdf (accessed on 1 February 2019).
6. Pojić, M.M.; Mastilović, J.S. Near Infrared Spectroscopy-Advanced Analytical Tool in Wheat Breeding, Trade, and Processing. *Food Biopro. Tech.* **2013**, *62*, 330–352. [CrossRef]
7. Shen, F.; Yang, D.; Ying, Y.; Li, B.; Zheng, Y.; Jiang, T. Discrimination Between Shaoxing Wines and Other Chinese Rice Wines by Near-Infrared Spectroscopy and Chemometrics. *Food Biopro. Tech.* **2012**, *52*, 786–795. [CrossRef]
8. Cortés, V.; Blasco, J.; Aleixos, N.; Cubero, S.; Talens, P. Visible and Near-Infrared Diffuse Reflectance Spectroscopy for Fast Qualitative and Quantitative Assessment of Nectarine Quality. *Food Biopro. Tech.* **2017**, *10*, 1755–1766. [CrossRef]

9. Magwaza, L.S.; Opara, U.L.; Nieuwoudt, H.; Cronje, P.J.R.; Saeys, W.; Nicolaï, B. FT-NIR Spectroscopy Applications for Internal and External Quality Analysis of Citrus Fruit-A Review. *Food Biopro. Tech.* **2012**, *52*, 425–444. [CrossRef]

10. Baca-Bocanegra, B.; Nogales-Bueno, J.; García-Estévez, I.; Escribano, T.; Hernandez-Hierro, J.M.; Heredia, F.J. Screening of Wine Extractable Total Phenolic and Ellagitannin Contents in Revalorized Cooperage By-products: Evaluation by Micro-FT-NIRS Technology. *Food Biopro. Tech.* **2019**, *12*, 477–485. [CrossRef]

11. Páscoa, R.N.M.J.; Machado, S.; Magalhães, L.M.; Lopes, J.A. Value Adding to Red Grape Pomace Exploiting Eco-friendly FT-NIR Spectroscopy Technique. *Food Biopro. Tech.* **2015**, *8*, 865–874. [CrossRef]

12. Parpinello, G.P.; Nunziatini, G.; Rombola, A.D.; Gottardi, F.; Versari, A. Relationship between sensory and FT-NIR spectroscopy in consumer preference of table grape (cv Italia). *Postharvest Biol. Tech.* **2013**, *83*, 47–53. [CrossRef]

13. Cane, P. Il controllo della qualità dei vini mediante HPLC: Determinazione degli acidi organici. *L'Enotecnico* **1990**, *26*, 69–72.

14. Association de coordination technique pour l'industrie agro-alimentaire (ACTIA). *Sensory Evaluation Guide of Good Practice*; Technical Report; Technical Coordination Association for the Food Industry: Paris, France, 2001; Available online: http://www.actia-asso.eu/cms/rubrique-2085-sensory_evaluation.html (accessed on 1 February 2019).

15. R Core Team. *R: A Language and Environment for Statistical Computing*; R Foundation for Statistical Computing: Vienna, Austria, 2013; Available online: http://www.R-project.org/ (accessed on 1 February 2019).

16. Siesler, H.W.; Ozaki, Y.; Kawata, S.; Heise, H.M. *Near Infrared Spectroscopy*; Wiley-VCH Verlags GmbH: Weinheim, Germany, 2002.

17. Martelo-Vidal, M.J.; Vázquez, M. Evaluation of Ultraviolet, Visible, and Near Infrared Spectroscopy for the Analysis of Wine Compounds. *Czech J. Food Sci.* **2014**, *32*, 37–47. [CrossRef]

18. Liu, F.; He, Y.; Wang, L.; Sun, G. Detection of organic acids and pH of fruit vinegars using near-infrared spectroscopy and multivariate calibration. *Food Biopro. Tech.* **2011**, *4*, 1331–1340. [CrossRef]

19. Da Costa Filho, P.A. Rapid determination of sucrose in chocolate mass using near infrared spectroscopy. *Anal. Chim. Acta* **2009**, *631*, 206–211. [CrossRef] [PubMed]

20. Ferrari, E.; Foca, G.; Vignali, M.; Tassi, L.; Ulrici, A. Adulteration of the anthocyanin content of red wines: Perspectives for authentication by Fourier Transform Near Infra Red and 1H NMR spectroscopies. *Anal. Chim. Acta* **2011**, *701*, 139–151. [CrossRef] [PubMed]

21. Simeone, M.L.F.; Parrella, R.A.C.; Schaffert, R.E.; Damasceno, C.M.B.; Leal, M.C.B.; Pasquini, C. Near infrared spectroscopy determination of sucrose, glucose and fructose in sweet sorghum juice. *Microchem. J.* **2017**, *134*, 125–130. [CrossRef]

22. Musingarabwi, D.M.; Nieuwoudt, H.H.; Young, P.R.; Eyeghe-Bickong, H.A.; Vivier, M.A. A rapid qualitative and quantitative evaluation of grape berries at various stages of development using Fourier-transform infrared spectroscopy and multivariate data analysis. *Food Chem.* **2016**, *190*, 253–262. [CrossRef]

23. Le Moigne, M.; Maury, C.; Bertrand, D.; Jourjon, F. Sensory and instrumental characterisation of Cabernet Franc grapes according to ripening stages and growing location. *Food Qual. Pref.* **2008**, *19*, 220–231. [CrossRef]

24. Marsico, A.D.; Perniola, R.; Cardone, M.F.; Velenosi, M.; Antonacci, D.; Alba, V.; Basile, T. Study of the Influence of Different Yeast Strains on Red Wine Fermentation with FT-NIR Spectroscopy and Principal Component Analysis. *Multidiscip. Sci. J.* **2018**, *1*, 13. [CrossRef]

25. Acri, G.; Testagrossa, B.; Vermiglio, G. FT-NIR Analysis of Different Garlic Cultivars. *J. Food Meas. Character.* **2016**, *10*, 127–136. [CrossRef]

26. Conzen, J.-P. *Multivariate Calibration*, 3rd ed.; Bruker Optik GmbH: Ettlingen, Germany, 2014; ISBN 978-3-929431-13-1.

27. Cozzolino, D.; Cynkar, W.U.; Shah, N.; Smith, P. Multivariate data analysis applied to spectroscopy: Potential application to juice and fruit quality. *Food Res. Intern.* **2011**, *44*, 1888–1896. [CrossRef]

28. Rambla, F.J.; Garrigues, S.; De la Guardia, M. PLS-FT-NIR determination of total sugar, glucose, fructose and sucrose in aqueous solutions of fruit juices. *Anal. Chim. Acta* **1997**, *344*, 41–53. [CrossRef]

29. Liu, L.; Cozzolino, D.; Cynkar, W.U.; Dambergs, R.D.; Janik, L.; O'Neill, B.K.; Colby, C.B.; Gishen, M. Preliminary study on the application of visible-near infrared spectroscopy and chemometrics to classify Riesling wines from different countries. *Food Chem.* **2008**, *106*, 781–786. [CrossRef]

30. Guimarães, C.C.; Assis, C.; Simeone, M.L.F.; Sena, M.M. Use of near-infrared spectroscopy, partial least-squares, and ordered predictors selection to predict four quality parameters of sweet sorghum juice used to produce bioethanol. *Energy Fuels* **2016**, *30*, 4137–4144. [CrossRef]
31. Cozzolino, D.; Cynkar, W.U.; Shah, N.; Smith, P. Can spectroscopy geographically classify Sauvignon Blanc wines from Australia and New Zeland? *Food Chem.* **2011**, *126*, 673–678. [CrossRef]

Article

Validation of a Commercial Loop-Mediated Isothermal Amplification (LAMP) Assay for the Rapid Detection of *Anisakis* spp. DNA in Processed Fish Products

Gaetano Cammilleri [1,2,*], Vincenzo Ferrantelli [1], Andrea Pulvirenti [2], Chiara Drago [3],
Giuseppe Stampone [3], Gema Del Rocio Quintero Macias [3], Sandro Drago [3], Giuseppe Arcoleo [3],
Antonella Costa [1], Francesco Geraci [1] and Calogero Di Bella [1]

[1]   Istituto Zooprofilattico Sperimentale della Sicilia, via Gino Marinuzzi 3, 90129 Palermo, Italy;
      vincenzo.ferrantelli@izssicilia.it (V.F.); antonella.costa@izssicilia.it (A.C.); francesco.geraci@izssicilia.it (F.G.);
      calogero.dibella@izssicilia.it (C.D.B.)
[2]   Dipartimento di Scienze della Vita, Università degli studi di Modena e Reggio Emilia, Via Università 4,
      41121 Modena, Italy; andrea.pulvirenti@unimore.it
[3]   Enbiotech s.r.l. Via Aquileia 34, 90144 Palermo, Italy; c.drago@enbiotech.eu (C.D.);
      g.stampone@enbiotech.eu (G.S.); g.delrocioquintero@enbiotech.eu (G.D.R.Q.M.);
      s.drago@enbiotech.eu (S.D.); g.arcoleo@enbiotech.eu (G.A.)
*   Correspondence: Gaetano.cammilleri86@gmail.com; Tel.: +39-328-8048262

Received: 27 November 2019; Accepted: 9 January 2020; Published: 16 January 2020

Abstract: Parasites belonging to the *Anisakis* genera are organisms of interest for human health because they are responsible for the Anisakiasis zoonosis, caused by the ingestion of raw or undercooked fish. Furthermore, several authors have reported this parasite to be a relevant inducer of acute or chronic allergic diseases. In this work, a rapid commercial system based on Loop-Mediated Isothermal Amplification (LAMP) was optimised and validated for the sensitive and rapid detection of *Anisakis* spp. DNA in processed fish products. The specificity and sensitivity of the LAMP assay for processed fish samples experimentally infected with *Anisakis* spp. larvae and DNA were determined. The LAMP system proposed in this study was able to give positive amplification for all the processed fish samples artificially contaminated with *Anisakis* spp., giving sensitivity values equal to 100%. Specificity tests provided no amplification for the *Contracaecum*, *Pseudoterranova*, or *Hysterothylacium* genera and uninfected samples. The limit of detection (LOD) of the LAMP assay proposed was 10^2 times lower than the real-time PCR method compared. To the best of our knowledge, this is the first report regarding the application of the LAMP assay for the detection of *Anisakis* spp. in processed fish products. The results obtained indicate that the LAMP assay validated in this work could be a reliable, easy-to-use, and convenient tool for the rapid detection of *Anisakis* DNA in fish product inspection.

Keywords: *Anisakis* spp.; molecular methods; LAMP; validation; anisakidae family

---

1. Introduction

The Anisakidae family includes a vast number of parasites with a worldwide distribution. The life-cycle of Anisakidae nematodes involves invertebrates, fish, cephalopods, and marine mammals so these parasites can be found in the muscles and viscera of numerous fish and cephalopod species [1–5]. In the Mediterranean, the *Anisakis* spp. parasites can be found in different teleosts belonging to different ecological distributions [6]. The fish species mainly involved in the life cycle of *Anisakis* belong to the pelagic, benthopelagic, and benthodemersal domains. Indeed, the three *Anisakis* species are widely occurring in pelagic, benthopelagic, and demersal species of the Gadidae, Merlucciidae, Scombridae,

Carangidae, and Trichiuridae families [2,7,8]. *Anisakis* spp. parasites are marine nematodes of health interest because of their high zoonotic potential, being responsible for a human disease called Anisakiasis. Anisakiasis is a zoonotic disease caused by the ingestion of raw or undercooked fish infected with Anisakidae larvae [9–13]. Anisakiasis has become an increasing human health concern, particularly in Asian countries, representing more than 50% of all cases, where the consumption of raw or undercooked fish is frequent and/or has become increasingly popular. The remaining cases are from European countries, especially Italy (57 cases from 1996 until now), Spain (124 cases from 1996 until now), and France (15 cases from 1975 until now), whereas there are much fewer cases in Scandinavian countries, despite comparatively high fish consumption rates per capita in these countries [7,14]. Moreover, several authors have reported this parasite to be a relevant inducer of acute or chronic allergic diseases [15–18]. *Anisakis* is implicated in allergic IgE-mediated reactions, such as urticaria, angioedema, asthma, and anaphylaxis, in highly sensitized people [19,20]. The European Food Safety Authority (EFSA 2010) confirmed that all wild saltwater fish must be considered at risk of containing viable parasites of human health concern and no sea fishing grounds can be considered free of *A. simplex complex* larvae [21].

EFSA also recommends further studies and methods to improve the surveillance and diagnostic awareness of pathologies to parasites in fishery products. Even the Italian Ministry of Health encourages fish sector operators to carry out correct evisceration protocols of fish products in order to prevent *Anisakis*-related pathologies by reducing the possibility of migration of L3 larvae in the musculature [22]. At present, European authorities perform laborious and unreliable inspection methods, such as visual control and transillumination with UV [23], which cannot be applied for processed fish products such as anchovy paste, marinated anchovies, infant formulas, etc. [24]. Furthermore, the current immunological methods for *Anisakis* allergy diagnosis give a high number of false positives due to the cross-reactivity with numerous panallergens [25,26].

In this case, molecular biology methods are valuable tools in the detection of *Anisakis* spp. nematodes in processed seafood [24,27–30]. Several studies have shown that immunological and molecular methodologies yield comparable results concerning the detection of allergens in processed foods as sensitive and specific tools [31–33]. According to literature, Real-Time PCR is recognized to be the only molecular method capable of detecting the presence of *Anisakis* spp. in processed fish products, showing satisfactory sensitivity and specificity values [24,27,30,33,34]. Nonetheless, this sophisticated and expensive molecular method is currently indispensable and requires operational skills limiting their broad applicability. Loop-mediated isothermal amplification (LAMP) is considered a highly sensitive and rapid method for DNA amplification at constant temperature (60–65 °C) [35,36].

The food sector operators must have systems and procedures that allow competent authorities to access information on the products in order to guarantee their hygiene. In this work, a commercial LAMP assay for *Anisakis* spp. DNA detection was optimised and validated in order to obtain a simple, fast, and cheap tool, which can identify possible risks to consumer health due to the presence of these organisms in processed fish products.

2. Materials and Methods

2.1. Fish Samples and Anisakis Larvae Collection

All the processed fish samples used for the method optimisation and validation came from a large-scale distribution, in order to reduce any bias from local food specialities and extend the range of validation. Homogenised farmed trout (*Oncorhynchus mykiss*; *n* = 40), homogenised farmed sea bream (*Saprus aurata*; *n* = 40), and homogenised farmed salmon (*Salmo salar*; *n* = 40) were chosen as naturally negative (noncontaminated by *Anisakis*) samples [4,21,37–39]; whereas anchovy (*Engraulis encreasicolus*) paste (*n* = 40), anchovy in oil (*n* = 40), and salted sardines (*Sardina pilchardus*; *n* = 40) samples were chosen as positive samples for the validation of the method and for matrix effects evaluation. All the processed fish samples came from Italian supermarkets. The Anisakidae larvae used for the artificial contamination of the samples were collected from *Lepidopus caudatus*, *Clupea harengus*, and *Merluccius merluccius* samples after visual inspection and modified chloro-peptic digestion [40]. The larvae isolated were washed in physiological saline serum (pH 7) and morphologically identified by B-800 light microscopy (Optika, Ponteranica, Italy) according to taxonomic keys [41]. Furthermore, *Anisakis* morphotype II, *Hysterothylacium* sp., *Contracaecum* sp., and *Pseudoterranova* sp. larvae were taken from the reference materials of the Centro di Referenza Nazionale per le Anisakiasi (Palermo, Italy). The number of *Anisakis* spp. larvae used for artificial contamination has been defined according to the prevalence of infestation of the fish species examined and reported in the literature [8,42–44]. The larvae were cut into small pieces and then were carefully mixed with the processed fish samples.

2.2. DNA Extraction

Genomic DNA was extracted from positive and negative fish samples, contaminated or not with *Anisakis* spp., respectively. The extraction was also carried out for the samples artificially infested with *Contracaecum* sp., *Pseudoterranova* sp., and *Hysterothylacium* sp. The DNA extraction was performed using a ready-to-use buffer contained in the Anisakis Screen Glow kit (Enbiotech S.r.l., Palermo, Italy). Then, 250 ± 50 mg of sample was directly placed into 15 mL tubes containing 4 mL of the ready-to-use extraction buffer (Enbiotech S.r.l., Palermo, Italy) and then incubated for 40 ± 5 min at room temperature.

2.3. Primers Design and LAMP Assays

To design the primer set targeting *Anisakis* spp. gene, the genomic sequences of the internal transcribed spacer 2 gene from various species were collected from GenBank™ (EU624342.1, AY826720.1, AB277823.1, AB196671.1, AB277821.1, AB551660.1, HF911524.1, AY826723.1, EU718479.1, JQ912690.1, KF512840.1, JX535521.1, KF032062.1, EU327691.1). A set of six primers, two outer (F3 and B3), two inner (FIP and BIP), and two loop (LF and LB), which recognised eight distinct regions of the target gene, was designed.

The analytical and diagnostic assays to recognise *Anisakis* spp. DNA was performed using Anisakis Screen Glow commercial kit (Enbiotech Group S.r.l., Palermo, Italy) with ICGENE mini portable instrument (Figure 1) (Enbiotech Group s.r.l., Palermo, Italy), consisting of a real-time fluorimeter, monitored and regulated by the the ICGENE application (Enbiotech Group s.r.l., Palermo, Italy), downloadable on various smart devices. The Anisakis Screen Glow commercial kit includes ready-to-use reaction tubes (containing primers, fluorescent dye, etc.) to achieve a rapid amplification of DNA template. The protocol to obtain the specific amplification of the target *Anisakis* spp. DNA was carried out in a mixture of a final volume of 55 µL, including 22 µL of Anisakis Screen Glow LAMP mix (Enbiotech Group s.r.l., Palermo, Italy), 30 µL of mineral oil, and 3 µL of the extracted DNA samples. The mineral oil was added to the top of the reaction mixture to prevent evaporation. The amplification was optimised and performed at 65 °C for 35 min. Real-time monitoring of the fluorescence associated with the amplification was possible using the fluorimeter of the ICGENE portable instrument and the ICGENE application interface.

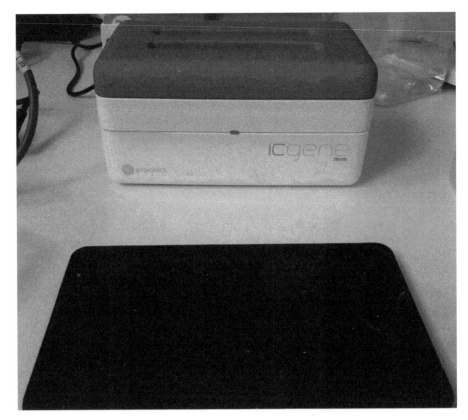

Figure 1. ICGENE mini portable instrument.

2.4. Specificity of the LAMP Assay

Based on the evolutionary relationships and their feasible genetic similarity, parasitic material belonging to the Anisakidae and Raphidascaridadae family was screened by the method proposed to have evidence on the diagnostic specificity of the LAMP assay. The parasitic material was previously characterised as belonging to *Hysterothylacium*, *Contracaecum*, and *Pseudoterranova* genera using specific molecular diagnostic keys and methods reported in the literature [45–47]. Two different experiments were carried out for the specificity evaluation: (i) In the first experiment performed using genomic DNA, *P. decipiens* sensu stricto, *P. krabbei*, *P. cattani*, *P. azarasi*, *C. rudolphii* A, and *Hysterothylacium aduncum* were tested in duplicate, also including internal positive and negative controls of the Anisakis Screen Glow commercial kit (Enbiotech S.r.l., Palermo, Italy); (ii) in the second experiment, processed fish samples artificially contaminated with *Pseudoterranova* sp., *Contracaecum* sp., and *Hysterothylacium* sp. larvae were subjected to DNA extraction and analysed in duplicate, together with positive and negative internal controls. Twenty samples of each processed fish product type, of which ten were artificially infested (5 with genomic DNA and 5 with larvae), were tested for the specificity assessment.

2.5. Sensitivity of the LAMP Assay

The sensitivity or inclusivity [48] was established as the ability of the LAMP method to detect DNA of *Anisakis* spp. (expressed as a percentage). Two different experiments were carried out for the inclusivity evaluation: (i) In the first experiment performed using genomic DNA, *A. pegreffii*, *A. simplex* sensu stricto, *A. typica*, *A.ziphidarum*, and *A. physeteris* were tested in duplicate, also including internal positive and negative controls; (ii) in the second experiment, processed fish samples artificially contaminated with *Anisakis* spp. type I and type II larvae were subjected to DNA extraction and analysed in duplicate, together with positive and negative internal controls. Twenty samples of each processed fish product type were tested for the inclusivity assessment.

2.6. Limit of Detection

The limit of detection (LOD) was established as the lowest concentration of DNA of *Anisakis* species, which provided a significantly different signal to the negative control. The determination of the LOD of the LAMP method was assessed by serial 10-fold dilution of the DNA extracted from *Anisakis* spp. larvae with nuclease-free water. All measurements were performed in ten replicates from each sample type independently. The range of the DNA extracted varied between 2.22 ng μL^{-1} and 8.40 ng μL^{-1} with good 260/280 and 260/230 ratios (1.8 to 2.1). A test was considered acceptable when it ensured the detection of positive samples successfully with DNA content equal to or greater than LOD.

2.7. Real-Time PCR Assay

Real-Time PCR (RT-PCR) assays of the same samples analysed by LAMP were also carried out for a comparative purpose and as a confirmation method for the results of the LAMP assays. The RT-PCR amplification was carried out using the materials and following the validated protocols described by Cavallero et al. (2017) [27]. The extraction of DNA from artificially or noncontaminated processed fish samples was carried out using the Ion Force Fast kit (Generon, Modena, Italy), following the manufacturer's instructions. The PCR reactions were carried out by the commercial kit PATHfinder Anisakis/Pseudoterranova DNA detection assay (Generon, Modena, Italy). Five microliters of DNA extracted were mixed with 15 µL of PATHfinder Anisakis/Pseudoterranova kit for a total volume of 20 µL. The RT-PCR was carried out in a BIOER 9600 series Thermocycler (BIOER, Hangzhou, China) following the present thermal cycling conditions: A Taq activation at 95 °C for 3 min, and 45 cycles of amplification (95 °C for 10 s and 57 °C for 16 s).

3. Results

3.1. Optimization of the LAMP Assay

The method was optimised for DNA extraction phase by testing in triplicate the initial weight of the samples at 50, 100, 250, and 350 mg with 0.5, 2, 4, and 8 mL of extraction buffer. The extract was tested with undiluted and diluted 1:5 and 1:10 for matrix effect assessment. An initial weight of 250 mg with 4 mL of extraction buffer and a 1:5 dilution after the extraction were found to be the best conditions for effective real-time detection of DNA amplification by the LAMP method proposed, giving the fluorescence intensity required for detection. The method was optimised by real-time monitoring of the time and temperature of reaction. The optimal reaction temperature and time for the LAMP assay was proved to be 65 °C and 35 min, respectively.

3.2. Sensitivity and Specificity

The LAMP method proposed was able to amplify *Anisakis* spp. DNA from artificially infested fish samples, giving a sensitivity of 100% for each sample type analysed (Figure 2). Moreover, the method was also able to detect each sample contaminated with *A. simplex* s.s., *A. pegreffii*, *A. physeteris*, *A. ziphidarum*, and *A. typica* DNA. The assay detected *Anisakis spp.* DNA to a dilution of 10^{-4} (0.00022 ng μL^{-1}), giving an amplification for all the replicates, with fluorescence values necessary for detection.

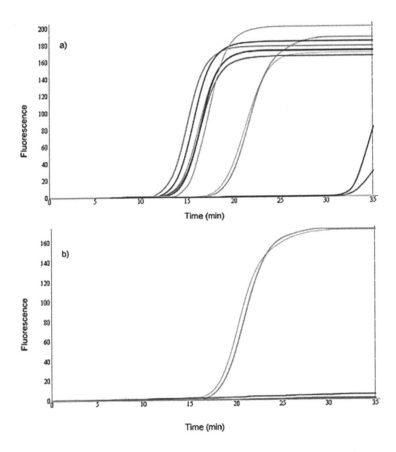

Figure 2. Monitoring of Loop-mediated isothermal amplification (LAMP) amplification for sensitivity (**a**) and specificity (**b**) tests from homogenized farmed salmon samples. The analysis shows the DNA amplification detection in real-time (colored curves shown in (**a**) and (**b**)) by measuring the increasing fluorescence of DNA binding to the dye. The amplification plots displayed in the specificity correspond to the positive control (duplicate).

All the LAMP analysis were carried out using positive and negative controls contained in the Anisakis Screen Glow kit. No amplification products were detected in uninfected samples, giving a very high specificity rate. Furthermore, no amplification was obtained on processed fish samples contaminated with *Contracaecum* sp., *Pseudoterranova* sp., and *Hysterothylacium* sp. larvae and DNA. The RT-PCR assay showed a sensitivity rate of 100% for the fish samples artificially infested with *Anisakis* sp. larvae and DNA. However, the analysis of uninfected samples and samples contaminated with *Pseudoterranova* sp., *Hysterothylacium* sp., and *Contracaecum* sp. larvae and DNA for the specificity test did not obtain any amplification.

In agarose gel analysis, the LAMP amplicons revealed a ladder-like pattern, according to what was reported in the literature, with many bands of different molecular weights, indicating the production of stem–loop DNA with inverted repeats of the target sequence (Parida et al., 2008; Li et al., 2012; Figure 3).

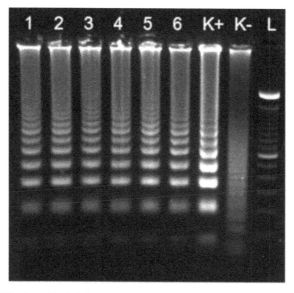

Figure 3. Amplification of LAMP for DNA extracted from anchovy paste samples experimentally infected with *Anisakis* spp. larvae (Lanes 1–6). Lane K+: Positive control, Lane K−: Negative control, Lane L: 100 bp DNA ladder.

4. Discussion

The method optimised with the set of primers employed was able to amplify *Anisakis* spp. DNA in 35 min at 65 °C from a considerable initial weight of samples, giving a satisfactory sensibility and specificity. The time needed for the LAMP assay was much lower than the RT-PCR method (about 94 min). The use of two loop primers and the ability of the two sets of primers (F3–B3, FIP–BIP) to recognise two distinct regions on the target DNA allowed acceleration of the reaction time and provided extremely high specificity. The specificity test has demonstrated that the LAMP assay proposed in this work can discriminate the *Anisakis* genera with respect to species belonging to other related genera that are not present in the Mediterranean Sea, such as *Pseudoterranova* spp. Furthermore, the method was able to discriminate the *Anisakis* genera with respect to nonallergic parasites such as *Hysterothylacium* spp, which proved that the LAMP primers are highly specific for the detection of *Anisakis* spp. The LAMP reaction primers recognise specifically eight independent regions in comparison with PCR primers, which can only recognise two independent regions, enhancing the sensitivity and specificity and decreasing the probability of false-positive results [49]. Generally, the fluorescence-based real-time monitoring of LAMP reaction is considerably faster than that performed by a real-time turbidimeter [50]. Moreover, compared to the real-time turbidity method, the real-time fluorescence method possesses two further advantages: The first is higher sensitivity; the second is that the sensitivity is less affected by the presence of opaque substances in the mixture, such as proteins [51]. The LAMP assay optimised was able to amplify *Anisakis* spp. DNA from different sample types, suggesting that the type of fish processing does not affect the quality of the assay by matrix interferences. The specificity and sensitivity of the LAMP method do not seem impaired by sample type, as confirmed by other studies [52]. However, there is a high risk of aerosol contamination due to the large amount of LAMP products. To reduce this risk, we adopted the use of mineral oil inside the reaction tubes.

The LAMP system we described here can detect the concentration of *Anisakis* spp. DNA up to 10^2 times lower than the RT-PCR method. The high sensitivity of such a method is undoubtedly an advantage, but needs special care to avoid false positives. The utilisation of ready-to-use reagents allowed us to minimise any operator error and the occurrence of possible false positives. Moreover, the presence of non-target DNA and inhibitors in the LAMP reaction was shown to not affect the amplification results. The present method in combination with the ICGENE mini portable instrument proved to be accurate as of the Real-Time PCR method, but more rapid, easy to use, and with a lower limit of detection. Furthermore, the LAMP system proposed in this work is inexpensive, requires little equipment and technical support, is not space-consuming, and is easy to use.

The IC GENE instrument allows one to analyze up to ten samples per test, excluding positive and negative controls, with very low costs. Given the satisfactory results obtained during the validation and the ease of use, the samples do not need to be analyzed in duplicate. *Anisakis* spp. nematodes are considered one of the most critical biological hazards present in seafood products [53].

These food-borne parasites are also a hidden food allergen [54]. Concerning the *A. simplex* complex, there is no information about the minimum allergen concentration that causes allergic diseases. In this respect, [34] concluded that a single larva contains sufficient allergens to induce an antibody response in sensitive individuals. Exposure to this parasite seems to increase due to the increasing consumption of seafood worldwide and the increase of new gastronomic trends based on the consumption of raw and undercooked fish. Because all the *Anisakis* spp. studied originate the same disease and allergic reactions, the species identification is less relevant [24]. To prevent the occurrence of this pathogen in fish sectors, effective and efficient control procedures must be adopted throughout the fish industry and processing continuum. Therefore, it is necessary to use more efficient, simple, and cost- and time-effective methods. Rapid or alternative methods are important, as they reduce the detection time for *Anisakis* spp. considerably in comparison to other inspective methods and prevent the dissemination of these nematodes along the food chain.

The assay validated in this work has the advantages of being more straightforward and more sensitive than PCR and real-time PCR [55]. The LAMP method also has the characteristics of not requiring special reagents and sophisticated temperature control devices so the detection of LAMP products is also suitable for on-site conditions. Furthemore, the preservation of the reagents used for the LAMP assay only requires storage at >4 °C. Therefore, the LAMP assay proposed in this work should be considered a new and reliable tool for food quality and security control for prevention of *Anisakis* allergy in fish sectors.

The present system could be suitable for use by operators of the fish product processing industry for self-monitoring purposes, given its extreme ease and speed, in order to verify the correct evisceration practices of fish sector operators, following the suggestion of the Italian Ministry of Health [22]. Furthermore, due to its capability of discriminating *Anisakis* spp. DNA with respect to species belonging to the *Pseudoterranova* genus, the LAMP system proposed in this work can be a valuable tool to trace Mediterranean fish products.

5. Conclusions

The results obtained confirmed the reliability of the method analysed, which is faster and has more sensibility (in terms of limit of detection) than the real-time PCR method compared. In conclusion, the great rapidity, sensitivity, and ease of use suggest that LAMP assay can be a valid alternative for routine examination in the fishery sector to help manufacturers establish concepts for hazard analysis and critical control points (HACCPs) evaluation, where sophisticated and expensive equipment is not sustainable.

Author Contributions: Conceptualisation, G.S. and G.C.; methodology, G.C., G.S., C.D., G.D.R.Q.M., and G.A.; validation, G.C. and F.G.; formal analysis G.S and G.C.; investigation, A.P. and S.D.; data curation, A.C., C.D.B., and G.C.; Writing—Original draft preparation, G.C. and G.S.; Writing—Review and editing A.P., V.F., and C.D.B; visualisation, V.F.; supervision, V.F. and C.D.B. All authors have read and agreed to the published version of the manuscript.

Funding: This research received no external funding.

Conflicts of Interest: The authors declare no conflict of interest.

References

1. Abollo, E.; Gestal, C.; Pascual, S. *Anisakis* infestation in marine fish and cephalopods from Galician waters: An updated perspective. *Parasitol. Res.* **2001**, *87*, 492–499. [PubMed]
2. Ferrantelli, V.; Costa, A.; Graci, S.; Buscemi, M.D.; Giangrosso, G.; Porcarello, C.; Palumbo, S.; Cammilleri, G. Anisakid Nematodes as Possible Markers to Trace Fish Products. *Ital. J. Food Saf.* **2015**, *4*, 4090. [CrossRef] [PubMed]
3. Mattiucci, S.; Nascetti, G. Chapter 2 Advances and Trends in the Molecular Systematics of Anisakid Nematodes, with Implications for Their Evolutionary Ecology and Host—Parasite Co-evolutionary Processes. In *Advances in Parasitology*; Academic Press: Cambridge, MA, USA, 2008; Volume 66, pp. 47–148. Available online: https://www.sciencedirect.com/science/article/abs/pii/S0065308X08002029 (accessed on 21 January 2017).
4. Cammilleri, G.; Costa, A.; Graci, S.; Buscemi, M.D.; Collura, R.; Vella, A.; Pulvirenti, A.; Cicero, A.; Giangrosso, G.; Schembri, P.; et al. Presence of *Anisakis pegreffii* in farmed sea bass (*Dicentrarchus labrax* L.) commercialized in Southern Italy: A first report. *Vet. Parasitol.* **2018**, *259*, 13–16. [CrossRef] [PubMed]
5. Chen, H.-X.; Zhang, L.-P.; Gibson, D.I.; Lü, L.; Xu, Z.; Li, H.-T.; Ju, H.-D.; Li, L. Detection of ascaridoid nematode parasites in the important marine food-fish *Conger myriaster* (Brevoort) (Anguilliformes: Congridae) from the Zhoushan Fishery, China. *Parasites Vectors* **2018**, *11*, 274. [CrossRef]
6. Gaglio, G.; Battaglia, P.; Costa, A.; Cavallaro, M.; Cammilleri, G.; Graci, S.; Buscemi, M.D.; Ferrantelli, V.; Andaloro, F.; Marino, F. *Anisakis* spp. larvae in three mesopelagic and bathypelagic fish species of the central Mediterranean Sea. *Parasitol. Int.* **2018**, *67*, 23–28. [CrossRef]
7. Mattiucci, S.; Cipriani, P.; Levsen, A.; Paoletti, M.; Nascetti, G. Chapter Four-Molecular Epidemiology of *Anisakis* and *Anisakiasis*: An Ecological and Evolutionary Road Map. In *Advances in Parasitology*; Rollinson, D., Stothard, J.R., Eds.; Academic Press: Cambridge, MA, USA, 2018; Volume 99, pp. 93–263.
8. Cammilleri, G.; Pulvirenti, A.; Costa, A.; Graci, S.; Collura, R.; Buscemi, M.D.; Sciortino, S.; Badaco, V.V.; Vazzana, M.; Brunone, M.; et al. Seasonal trend of Anisakidae infestation in South Mediterranean bluefish. *Nat. Prod. Res.* **2020**, *34*, 158–161. [CrossRef]
9. Arizono, N.; Yamada, M.; Tegoshi, T.; Yoshikawa, M. *Anisakis simplex* sensu stricto and *Anisakis pegreffii*: Biological Characteristics and Pathogenetic Potential in Human Anisakiasis. *Foodborne Pathog. Dis.* **2012**, *9*, 517–521. [CrossRef]
10. Bao, M.; Pierce, G.J.; Pascual, S.; González-Muñoz, M.; Mattiucci, S.; Mladineo, I.; Cipriani, P.; Bušelić, I.; Strachan, N.J.C. Assessing the risk of an emerging zoonosis of worldwide concern: Anisakiasis. *Sci. Rep.* **2017**, *7*, 43699. [CrossRef]
11. Fumarola, L.; Monno, R.; Ierardi, E.; Rizzo, G.; Giannelli, G.; Lalle, M.; Pozio, E. *Anisakis pegreffi* Etiological Agent of Gastric Infections in Two Italian Women. *Foodborne Pathog. Dis.* **2009**, *6*, 1157–1159. [CrossRef]
12. Polimeno, L.; Loiacono, M.; Pesetti, B.; Mallamaci, R.; Mastrodonato, M.; Azzarone, A.; Annoscia, E.; Gatti, F.; Amoruso, A.; Ventura, M.T. *Anisakiasis*, an Underestimated Infection: Effect on Intestinal Permeability of *Anisakis simplex*–Sensitized Patients. *Foodborne Pathog. Dis.* **2010**, *7*, 809–814. [CrossRef]
13. Sakanari, J.A.; McKerrow, J.H. Anisakiasis. *Clin. Microbiol. Rev.* **1989**, *2*, 278–284. [CrossRef]
14. Audicana, M.T.; Ansotegui, I.J.; de Corres, L.F.; Kennedy, M.W. *Anisakis simplex*: Dangerous—Dead and alive? *Trends Parasitol.* **2002**, *18*, 20–25. [CrossRef]
15. Alonso, A.; Daschner, A.; Moreno-Ancillo, A. Anaphylaxis with *Anisakis simplex* in the Gastric Mucosa. *N. Engl. J. Med.* **1997**, *337*, 350–351. [CrossRef] [PubMed]
16. Alonso-Gómez, A.; Moreno-Ancillo, A.; López-Serrano, M.C.; Suarez-de-Parga, J.M.; Daschner, A.; Caballero, M.T.; Barranco, P.; Cabañas, R. *Anisakis simplex* only provokes allergic symptoms when the worm parasitises the gastrointestinal tract. *Parasitol. Res.* **2004**, *93*, 378–384. [CrossRef] [PubMed]
17. Anadón, A.M.; Romarís, F.; Escalante, M.; Rodríguez, E.; Gárate, T.; Cuéllar, C.; Ubeira, F.M. The *Anisakis simplex* Ani s 7 major allergen as an indicator of true *Anisakis* infections. *Clin. Exp. Immunol.* **2009**, *156*, 471–478. [CrossRef] [PubMed]
18. Rodríguez-Mahillo, A.I.; González-Muñoz, M.; De las Heras, C.; Tejada, M.; Moneo, I. Quantification of *Anisakis simplex* Allergens in Fresh, Long-Term Frozen, and Cooked Fish Muscle. *Foodborne Pathog. Dis.* **2010**, *7*, 967–973. [CrossRef]
19. Choi, S.-J.; Lee, J.-C.; Kim, M.-J.; Hur, G.-Y.; Shin, S.-Y.; Park, H.-S. The Clinical Characteristics of *Anisakis* Allergy in Korea. *Korean J. Intern. Med.* **2009**, *24*, 160–163. [CrossRef]

20. Mazzucco, W.; Raia, D.D.; Marotta, C.; Costa, A.; Ferrantelli, V.; Vitale, F.; Casuccio, A. *Anisakis* sensitization in different population groups and public health impact: A systematic review. *PLoS ONE* **2018**, *13*, e0203671. [CrossRef]

21. European Food Safety Authority. Scientific Opinion on risk assessment of parasites in fishery products. *EFSA J.* **2010**, *8*, 1543. [CrossRef]

22. Salute, M. *Sicurezza Alimentare, Linee Guida in Materia di Igiene dei Prodotti della Pesca*; Ministero della Salute: Rome, Italy, 2015.

23. Seo, J.-S.; Jun, E.-J.; Jung, S.-H.; Kim, M.-S.; Park, M.-A.; Lee, C.-H.; Han, M.-C.; Kim, J.-W.; Jee, B.-Y. Prevalence of Anisakid larvae in chum salmon *Oncorhynchus keta* in Korea. *J. Fish Pathol.* **2010**, *23*, 123–129.

24. Herrero, B.; Vieites, J.M.; Espiñeira, M. Detection of anisakids in fish and seafood products by real-time PCR. *Food Control* **2011**, *22*, 933–939. [CrossRef]

25. Lluch-Bernal, M.; Sastre, J.; Fernández-Caldas, E.; Marañon, F.; Cuesta-Herranz, J.; las Heras De, M.; Quirce, S.; Novalbos, A. Conjunctival provocation tests in the diagnosis of *Anisakis simplex* hypersensitivity. *J. Investig. Allergol. Clin. Immunol.* **2002**, *12*, 21–24. [PubMed]

26. García-Palacios, L.; González, M.L.; Esteban, M.I.; Mirabent, E.; Perteguer, M.J.; Cuéllar, C. Enzyme-linked immunosorbent assay, immunoblot analysis and RAST fluoroimmunoassay analysis of serum responses against crude larval antigens of *Anisakis simplex* in a Spanish random population. *J. Helminthol.* **1996**, *70*, 281–289. [CrossRef] [PubMed]

27. Cavallero, S.; Bruno, A.; Arletti, E.; Caffara, M.; Fioravanti, M.L.; Costa, A.; Cammilleri, G.; Graci, S.; Ferrantelli, V.; D'Amelio, S. Validation of a commercial kit aimed to the detection of pathogenic anisakid nematodes in fish products. *Int. J. Food Microbiol.* **2017**, *257*, 75–79. [CrossRef]

28. Godínez-González, C.; Roca-Geronès, X.; Cancino-Faure, B.; Montoliu, I.; Fisa, R. Quantitative SYBR Green qPCR technique for the detection of the nematode parasite *Anisakis* in commercial fish-derived food. *Int. J. Food Microbiol.* **2017**, *261*, 89–94. [CrossRef] [PubMed]

29. Mossali, C.; Palermo, S.; Capra, E.; Piccolo, G.; Botti, S.; Bandi, C.; D'Amelio, S.; Giuffra, E. Sensitive Detection and Quantification of Anisakid Parasite Residues in Food Products. *Foodborne Pathog. Dis.* **2009**, *7*, 391–397. [CrossRef]

30. Zhang, S.L.; Wang, H.L.; Liu, J.; Ni, F.; Xu, S.S.; Luo, D.M. Detection of anisakid nematodes by an SYBR green I real-time PCR. *Zhongguo Ji Sheng Chong Xue Yu Ji Sheng Chong Bing Za Zhi* **2010**, *28*, 194–199.

31. Costa, J.; Ansari, P.; Mafra, I.; Oliveira, M.B.P.P.; Baumgartner, S. Assessing hazelnut allergens by protein- and DNA-based approaches: LC-MS/MS, ELISA and real-time PCR. *Anal. Bioanal. Chem.* **2014**, *406*, 2581–2590. [CrossRef] [PubMed]

32. Stephan, O.; Weisz, N.; Vieths, S.; Weiser, T.; Rabe, B.; Vatterott, W. Protein Quantification, Sandwich ELISA, and Real-Time PCR Used to Monitor Industrial Cleaning Procedures for Contamination with Peanut and Celery Allergens. *J. AOAC Int.* **2004**, *87*, 1448–1457. Available online: https://www.ingentaconnect.com/content/aoac/jaoac/2004/00000087/00000006/art00022 (accessed on 20 May 2019).

33. Stephan, O.; Vieths, S. Development of a Real-Time PCR and a Sandwich ELISA for Detection of Potentially Allergenic Trace Amounts of Peanut (*Arachis hypogaea*) in Processed Foods. *J. Agric. Food Chem.* **2004**, *52*, 3754–3760. [CrossRef]

34. Lopez, I.; Pardo, M.A. Evaluation of a Real-Time Polymerase Chain Reaction (PCR) Assay for Detection of *Anisakis simplex* Parasite as a Food-Borne Allergen Source in Seafood Products. *J. Agric. Food Chem.* **2010**, *58*, 1469–1477. [CrossRef]

35. Notomi, T.; Okayama, H.; Masubuchi, H.; Yonekawa, T.; Watanabe, K.; Amino, N.; Hase, T. Loop-mediated isothermal amplification of DNA. *Nucleic Acids Res.* **2000**, *28*, e63. [CrossRef] [PubMed]

36. Nagamine, K.; Hase, T.; Notomi, T. Accelerated reaction by loop-mediated isothermal amplification using loop primers. *Mol. Cell. Probes* **2002**, *16*, 223–229. [CrossRef] [PubMed]

37. Peñalver, J.; Dolores, E.M.; Muñoz, P. Absence of Anisakid Larvae in Farmed European Sea Bass (*Dicentrarchus labrax* L.) and Gilthead Sea Bream (*Sparus aurata* L.) in Southeast Spain. *J. Food Prot.* **2010**, *73*, 1332–1334.

38. Lunestad, B.T. Absence of Nematodes in Farmed Atlantic Salmon (*Salmo salar* L.) in Norway. *J. Food Prot.* **2003**, *66*, 122–124. [CrossRef]

39. Angot, V.; Brasseur, P. European farmed Atlantic salmon (*Salmo salar* L.) are safe from anisakid larvae. *Aquaculture* **1993**, *118*, 339–344. [CrossRef]

40. Cammilleri, G.; Chetta, M.; Costa, A.; Graci, S.; Collura, R.; Buscemi, M.D.; Cusimano, M.; Alongi, A.; Principato, D.; Giangrosso, G.; et al. Validation of the TrichinEasy® digestion system for the detection of *Anisakidae larvae* in fish products. *Acta Parasitol.* **2016**, *61*, 369–375. [CrossRef]
41. Berland, B. Nematodes from some Norwegian marine fishes. *Sarsia* **1961**, *2*, 1–50. [CrossRef]
42. Cavallero, S.; Magnabosco, C.; Civettini, M.; Boffo, L.; Mingarelli, G.; Buratti, P.; Giovanardi, O.; Fortuna, C.M.; Arcangeli, G. Survey of *Anisakis* sp. and *Hysterothylacium* sp. in sardines and anchovies from the North Adriatic Sea. *Int. J. Food Microbiol.* **2015**, *200*, 18–21. [CrossRef]
43. Ciccarelli, C.; Aliventi, A.; Trani, V.D.; Semeraro, A.M. Assessment of the prevalence of *Anisakidae larvae* prevalence in anchovies in the central Adriatic Sea. *Ital. J. Food Saf. 1(1zero)* **2011**, *1*, 15–19. [CrossRef]
44. Cipriani, P.; Acerra, V.; Bellisario, B.; Sbaraglia, G.L.; Cheleschi, R.; Nascetti, G.; Mattiucci, S. Larval migration of the zoonotic parasite *Anisakis pegreffii* (Nematoda: Anisakidae) in European anchovy, *Engraulis encrasicolus*: Implications to seafood safety. *Food Control* **2016**, *59*, 148–157. [CrossRef]
45. Cavallero, S.; Nadler, S.A.; Paggi, L.; Barros, N.B.; D'Amelio, S. Molecular characterization and phylogeny of anisakid nematodes from cetaceans from southeastern Atlantic coasts of USA, Gulf of Mexico, and Caribbean Sea. *Parasitol. Res.* **2011**, *108*, 781–792. [CrossRef] [PubMed]
46. Costa, A.; Cammilleri, G.; Graci, S.; Buscemi, M.D.; Vazzana, M.; Principato, D.; Giangrosso, G.; Ferrantelli, V. Survey on the presence of *A. simplex* s.s. and A. pegreffii hybrid forms in Central-Western Mediterranean Sea. *Parasitol. Int.* **2016**, *65*, 696–701. [CrossRef] [PubMed]
47. D'Amelio, S.; Mathiopoulos, K.D.; Santos, C.P.; Pugachev, O.N.; Webb, S.C.; Picanço, M.; Paggi, L. Genetic markers in ribosomal DNA for the identification of members of the genus *Anisakis* (Nematoda: Ascaridoidea) defined by polymerase-chain-reaction-based restriction fragment length polymorphism. *Int. J. Parasitol.* **2000**, *30*, 223–226. [CrossRef]
48. Feldsine, P.; Abeyta, C.; Andrews, W.H.; AOAC International Methods Committee. AOAC International methods committee guidelines for validation of qualitative and quantitative food microbiological official methods of analysis. *J. AOAC Int.* **2002**, *85*, 1187–1200. [PubMed]
49. Li, X.; Liu, W.; Wang, J.; Zou, D.; Wang, X.; Yang, Z.; Yin, Z.; Cui, Q.; Shang, W.; Li, H.; et al. Rapid detection of *Trichinella spiralis* larvae in muscles by loop-mediated isothermal amplification. *Int. J. Parasitol.* **2012**, *42*, 1119–1126. [CrossRef]
50. Zhang, X.; Lowe, S.B.; Gooding, J.J. Brief review of monitoring methods for loop-mediated isothermal amplification (LAMP). *Biosens. Bioelectron.* **2014**, *61*, 491–499. [CrossRef]
51. Francois, P.; Tangomo, M.; Hibbs, J.; Bonetti, E.-J.; Boehme, C.C.; Notomi, T.; Perkins, M.D.; Schrenzel, J. Robustness of a loop-mediated isothermal amplification reaction for diagnostic applications. *FEMS Immunol. Med. Microbiol.* **2011**, *62*, 41–48. [CrossRef]
52. Kong, Q.-M.; Lu, S.-H.; Tong, Q.-B.; Lou, D.; Chen, R.; Zheng, B.; Kumagai, T.; Wen, L.-Y.; Ohta, N.; Zhou, X.-N. Loop-mediated isothermal amplification (LAMP): Early detection of *Toxoplasma gondii* infection in mice. *Parasites Vectors* **2012**, *5*, 2. [CrossRef]
53. Scientific Opinion on Risk Assessment of Parasites in Fishery Products|Autorità Europea per la Sicurezza Alimentare. Available online: https://www.efsa.europa.eu/it/efsajournal/pub/1543 (accessed on 30 June 2017).
54. Nieuwenhuizen, N.E.; Lopata, A.L. Allergic Reactions to *Anisakis* Found in Fish. *Curr. Allergy Asthma Rep.* **2014**, *14*, 455. [CrossRef]
55. Parida, M.; Sannarangaiah, S.; Dash, P.K.; Rao, P.V.L.; Morita, K. Loop mediated isothermal amplification (LAMP): A new generation of innovative gene amplification technique; perspectives in clinical diagnosis of infectious diseases. *Rev. Med. Virol.* **2008**, *18*, 407–421. [CrossRef] [PubMed]

Article

A Real-Time PCR Screening Assay for Rapid Detection of *Listeria Monocytogenes* Outbreak Strains

Marina Torresi [1], Anna Ruolo [1], Vicdalia Aniela Acciari [1], Massimo Ancora [1], Giuliana Blasi [2], Cesare Cammà [1], Patrizia Centorame [1], Gabriella Centorotola [1], Valentina Curini [1], Fabrizia Guidi [2], Maurilia Marcacci [1], Massimiliano Orsini [3], Francesco Pomilio [1] and Marco Di Domenico [1,*]

[1]   Istituto Zooprofilattico Sperimentale dell'Abruzzo e del Molise G. Caporale, via Campo Boario, 64100 Teramo TE, Italy; m.torresi@izs.it (M.T.); anna.ruolo@libero.it (A.R.); v.acciari@izs.it (V.A.A.); m.ancora@izs.it (M.A.); c.camma@izs.it (C.C.); p.centorame@izs.it (P.C.); g.centorotola@izs.it (G.C.); v.curini@izs.it (V.C.); m.marcacci@izs.it (M.M.); f.pomilio@izs.it (F.P.)
[2]   Istituto Zooprofilattico Sperimentale dell'Umbria e delle Marche Togo Rosati, Via Gaetano Salvemini, 1, 06126 Perugia PG, Italy; g.blasi@izsum.it (G.B.); f.guidi@izsum.it (F.G.)
[3]   Istituto Zooprofilattico Sperimentale delle Venezie, Viale dell'Università, 10, 35020 Legnaro PD, Italy; morsini@izsvenezie.it
*   Correspondence: m.didomenico@izs.it

Received: 6 December 2019; Accepted: 5 January 2020; Published: 8 January 2020

Abstract: From January 2015 to March 2016, an outbreak of 23 human cases of listeriosis in the Marche region and one human case in the Umbria region of Italy was caused by *Listeria monocytogenes* strains showing a new pulsotype never described before in Italy. A total of 37 clinical strains isolated from patients exhibiting listeriosis symptoms and 1374 strains correlated to the outbreak were received by the Italian National Reference Laboratory for *L. monocytogenes* (It NRL Lm) of Istituto Zooprofilattico Sperimentale dell'Abruzzo e del Molise (IZSAM) for outbreak investigation. A real-time PCR assay was purposely designed for a rapid screening of the strains related to the outbreak. PCR-positive strains were successively typed through molecular serogrouping, pulsed field gel electrophoresis (PFGE), and Next Generation Sequencing (NGS). Applying the described strategy, based on real-time PCR screening, we were able to considerably reduce time and costs during the outbreak investigation activities.

Keywords: *Listeria monocytogenes*; outbreak; molecular methods; real-time PCR; screening

---

1. Introduction

Listeriosis is a foodborne illness caused by *Listeria monocytogenes*. The most common symptoms of the mild form of the disease include diarrhoea, fever, headache, and myalgia. However, when listeriosis appears as an invasive infection, patients may develop more severe outcomes such as meningitis and/or septicemia in adults, infection of the foetus and miscarriage in pregnant women, or neonatal infection [1]. The disease is also associated with a high mortality rate, reaching 20%–30%, and for risk-group patients even 75%. The number of confirmed cases of listeriosis among the inhabitants of the EU demonstrated a growing tendency from 1439 reported in 2005 to 2549 in 2018 [2,3].

Consistent with the increasing trend at European level, an intensification in the occurrence of listeriosis was observed in Marche region, Italy, between January 2015 and May 2015. After that, a total of 24 human cases of listeriosis occurred [4].

During the outbreak a huge number of strains were sent to the Italian National Reference Laboratory for *L. monocytogenes* (It NRL Lm) in order to identify the source of infection, and to carry out trace back and forward activities.

Next generation sequencing of the isolates and comparative genome analysis confirmed a unique strain responsible for the outbreak.

Since 2000, pulsed field gel electrophoresis (PFGE) has become the gold standard for *L. monocytogenes* subtyping and has been extensively used, throughout the world, during outbreak and trace back investigations. Although PFGE is widely accepted and used, in some cases it was found to be inadequate and obsolete [5]. Nowadays, the use of whole genome sequencing (WGS) for the characterization of pathogens has become a standard component of infectious disease surveillance and WGS-based differentiation of *L. monocytogenes* isolates has become a pivotal tool for listeriosis outbreak investigations in the USA [6,7]. Recent advances in sequencing technologies and analysis tools have rapidly increased the output and the analysis speed and also reduced the costs of WGS [8].

Despite this, PFGE, thanks to its internationally accredited and standardized protocol, is often joined with the Single Nucleotide Polymorphism (SNP)-based analyses [6]. In a hypothetical scenario, when outbreak occurs, thousands of samples could be processed and a huge number of *L. monocytogenes* strains can be isolated from human, food processing environment, and food sources leading to a possible bottleneck in both sequencing/data analysis and PFGE typing. This protocol requires up to 40 working days to accomplish strain typing. The response time of the analyses performed can make the difference between the occurrence of new clinical cases or not, and between minor or major economic loss for the food business operators and farms involved. For all these reasons the use of a screening method is a focal point for proficient outbreak management.

The aim of this paper was to describe the application of a specific real-time PCR screening assay properly designed for the rapid identification of strains potentially related to the listeriosis outbreak occurred in Central Italy between 2015 and 2016.

2. Materials and Methods

2.1. Bacterial Isolates DNA Extraction

All the strains were received as pure culture on sheep blood agar and subjected to DNA extraction. Stock cultures (Microbank™, Pro Lab Diagnostics Inc., Richmond Hill, ON, Canada) were prepared and stored at −80 °C when the strains were not processed immediately.

Cultures identified as *L. monocytogenes* by biochemical methods were grown overnight in sheep blood agar (Microbiol & C. s.n.c., Cagliari, Italy), picked, and dissolved in 300 µL of nuclease-free water (Ambion™, Thermo Fisher Scientific, Waltham, MA, USA). Then, 100 µL of 20 mg/mL lysozyme was added and incubated for 2 h at 56 °C. Finally, 300 µL of the suspension were transferred to the cartridges provided by the Maxwell 16 Cell DNA Purification Kit (Promega, Madison, WI, USA). DNA extraction was accomplished following the manufacturer's instructions. DNA was quantified using a Qubit dsDNA HS (High Sensitivity) Assay Kit (Invitrogen, Carlsbad, CA, USA) and purity was checked by a Nanodrop ND-1000 spectrophotometer (Thermo Fisher Scientific, Waltham, MA, USA).

2.2. Serogroup

Strain characterization was performed with a molecular serogroup-related PCR completed by the detection of *flaA* [9,10].

Briefly, DNA was extracted as described before and reactions were carried out in a Gene Amp PCR System 9700 thermal cycler (Applied Biosystems, Foster City, CA, USA). PCR products were run in 2% (*w/v*) agarose gel in 1× tris/borate/EDTA (TBE) buffer (Biorad, Hercules, CA, USA) and visualized by SYBR™ Safe DNA Gel Stain (Thermo Fisher Scientific, Waltham, MA, USA).

2.3. PFGE

PFGE analysis was performed, according to PulseNet protocol [11], as described previously [12]. Briefly, the bacterial suspensions were embedded in agarose, lysed, washed, and then digested with the restriction enzymes. The digested samples underwent electrophoresis in 1% (*w/v*) SeaKem Gold agarose (Lonza Rockland, Inc., Rockland, ME, USA) in 1 × TBE (Sigma-Aldrich, St. Louis, MO, USA) by using the Chef MapperXA system (Bio-Rad Laboratories, Hercules, CA, USA) at 6 V/cm, with a pulse time between 4 and 40 s for 19 h.

PFGE profiles were analyzed using BioNumerics version 7.5 (Applied Maths, Sint-Martens-Latem, Belgium). The similarities between the *Asc*I and *Apa*I macrorestriction profiles (MRPs) were calculated using the Dice coefficient, applying an optimization coefficient and band tolerance of 1.0% for both enzymes.

2.4. Next Generation Sequencing

WGS was used, in the first instance, to find common genomic regions for developing a real-time PCR screening test and then to perform outbreak inclusion/exclusions by SNPs analysis.

The DNA from *L. monocytogenes* strains were sequenced by the NextSeq500 Illumina platform using the Nextera XT protocol. Raw data were trimmed and assembled using Trimmomatic [13] and Spades 3.11 [14], respectively, with default parameters for the Illumina platform 2 × 150 chemistry. Then the genomes annotation was performed by Prokka [15] with default parameters except for the bacterial genetic code (−gcode 11). Annotation data were used to build a pan-genome matrix using roary [16] with default parameters and the .gff files obtained by the annotation.

Genetic relationships among isolates and outbreak inclusion/exclusions were performed by a SNP-based approach, using the reference free tool, kSNP3 [17], and a kmer value of 21 as indicated by Morganti et al. [18]. The core SNPs matrix was used as input to draw a neighbor-joining (NJ) phylogenetic tree using Mega6 [19].

2.5. Real-Time PCR

Aligning whole genome sequences of 12 outbreak clinical strains by ClustalW, common genomic sequences were identified [20]. Multiple sequence alignment found 14 highly conserved regions (coding for hypothetical proteins, recombinase family protein, transcriptional regulator, DEAD-DEAH box helicase, GNAT family acetyltransferase, and N-6 DNA methylase).

Primer Express v3.0.1 software (Applied Biosystems, Foster City, CA, USA) was used to design TaqMan assays for all the 14 genome regions. The recombinase family protein gene (rec) and the transcriptional regulator (trans) assays, which showed the best score, were selected to develop a multiplex real-time PCR screening method. The assays were optimized in a duplex real-time PCR using 6-Carboxyfluorescein (FAM) and 6-Carboxy-4',5'-Dichloro-2',7'-Dimethoxyfluorescein (JOE) as fluorescent reporter dyes, respectively (Table 1).

Table 1. Primers and probes sequences for Rec and Trans real-time PCR.

Oligonucleotide	Sequence 5'-3'	Size
Rec-fwd	AAATAATGCGGAGTTAAAAGGTGAA	
Rec-rev	TGGACTGCATTTGGTATGTGAGT	74 bp
Rec-probe	FAM-TACGGATTGCCGTCCCCGAAAGT-BHQ1	
Trans-fwd	CTCATTACGTTGATTGGCATACG	
Trans-rev	GGTTCGTGGTCTCCTTTTACAATAA	79 bp
Trans-probe	JOE-AACGAAGAAAAGGGAAAAACTCCCACCC-BHQ1	

Oligonucleotides were synthesized by Eurofins Genomics (Ebersberg, Germany). The 20 μL reaction volume contained 5 μL of purified DNA (2 ng/μL), 10 μL of GoTaq Probe qPCR Master Mix 2×, 300 nM for both Rec forward and reverse primers, 150 nM for Rec probe, 600 nM for both Trans forward and reverse primers, 200 nM for Trans probe, and nuclease-free water up to final volume. Real-time PCR were performed on the 7900 HT Fast Real-Time PCR System (Applied Biosystems, Foster City, CA, USA) using the following thermal profile: DNA polymerase activation at 95 °C for 20 s tailed by 35 cycles of denaturation at 95 °C for 1 s and annealing/extension at 60 °C for 20 s.

Three replicates of five ten-fold DNA serial dilutions from 20 to 0.002 ng/μL were amplified to create the standard curve. The efficiency (E) was calculated according to the formula $E = (10^{-1/slope} - 1) \times 100$ [21].

3. Results

Between January 2015 and September 2016, a total number of 37 clinical strains isolated from patients exhibiting listeriosis symptoms and 1374 strains correlated to the outbreak were received by the It NRL Lm. Among them, 1397 were screened with real-time PCR, 1230 were typed with molecular serogroup-related PCR, and 490 with PFGE and WGS.

3.1. Workflow

Real-time PCR and serogroup determination were used as alternative screening methods to give priority for deeper typing by Next Generation Sequencing (NGS) and PFGE. Before real-time PCR, strain selection was based on serogroup determination and then PFGE. The robustness coupled with the reduced turnaround time for analysis made the real-time PCR the method of choice for screening purposes. For every sample, up to five colonies were screened and then one PCR-positive colony was sequenced by NGS and genotyped by SNPs analysis (Figure 1). All the colonies sequenced were also genotyped by PFGE.

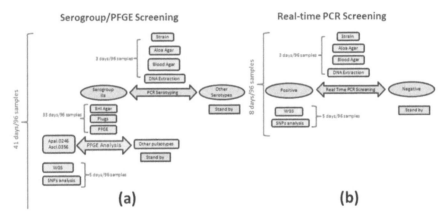

Figure 1. Analysis workflow and turnaround time to analyze 96 samples. Cases (**a**) and (**b**) show time of analysis applying serogroup and PFGE and real-time PCR as screening test to perform outbreak inclusion/exclusion, respectively.

3.2. Serogrouping and PFGE

Serogrouping analysis clustered 1224 isolates into four serogroups (IIa flaA+, IIc flaA−, IVb, and IIb), one strain was associated to serogroup IIc flaA+, and five strains were identified as *Listeria* spp.

The main serogroup was IIa flaA+ (47.7% $n = 587$). Serogroup IIc flaA− was detected in 28.4% of the strains ($n = 349$), while serogroups IIb and IVb were identified in 13.3% ($n = 164$) and 10.1% ($n = 124$), respectively.

All the strains included in the outbreak showed the same serogroup, IIa flaA+, and pulsotype *Apa*I.0246 *Asc*I.0356 as defined by the European Centre for Disease Prevention and Control (ECDC).

The PFGE analysis was carried out on 490 strains with both the enzymes and yielded 97 combined pulsotypes in addition to the outbreak profile that represented the prevalent one (21.6%) (Figure 2).

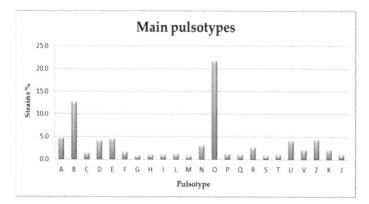

Figure 2. Main pulsotypes detected in 490 *Listeria monocytogenes* strains analyzed during the outbreak investigation.

3.3. Next Generation Sequencing

DNA of real-time PCR positive strains was sequenced by NGS and then related strains typed by PFGE. The theoretical coverage for each sample was higher than 70× with an average reads quality higher than 32.

Genetic relationships among isolates were evaluated through a SNP-based approach. The threshold for outbreak inclusion was set to 60 different SNPs. The neighbor joining phylogenetic tree showed a strong correlation between strains included in the outbreak, suggesting a high clonality of the collected strains (Figure 3).

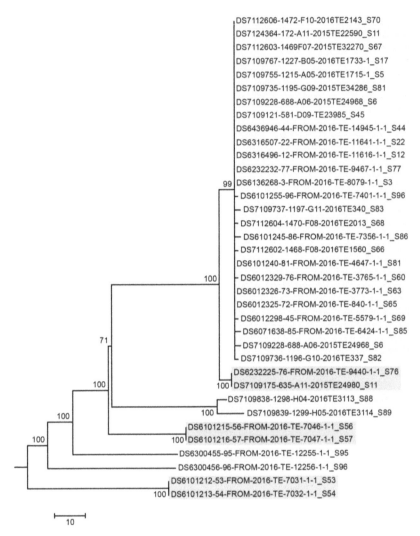

Figure 3. Phylogenetic relationships (neighbor joining (NJ)) among a selected subset of 38 strains analysed during the outbreak investigation. Strains showing pulsotype *ApaI.0246 AscI.0356* are highlighted in yellow; strains positive to real-time PCR but with a different pulsotype from the outbreak strain are highlighted in purple.

3.4. Real-Time PCR

Analytical specificity, the ability of the assay to distinguish the DNA target from non-target DNA, was assessed on 639 strains not included in the outbreak based on SNP analysis, while 111 strains, included in the epidemic cluster, were used as positive controls.

The efficiencies of real-time PCR assays for the target Rec and Trans were 99% and 94%, respectively (Table 2). All the 111 strains previously included in the outbreak by SNP analysis were correctly detected by both assays, while the DNA of the 639 strains not included in the outbreak were not amplified. The analytical specificity calculated was 100% (lower confidence limit 99.5% and upper confidence limit 100%).

Table 2. Analytical performance of the duplex real-time PCR method.

Assay	Slope	R^2	Efficiency
Rec	−3.34	0.998	99%
Trans	−3.47	0.999	94%

Efficiency $= (10^{-1/\text{slope}} - 1) \times 100$.

The method was then performed on 1397 strains during the outbreak investigation. A total of 1178 (84.3%) strains turned out to be negative, while 219 (15.7%) were detected by both targets. All positive samples were immediately sequenced by NGS and typed by molecular serogroup-related PCR and PFGE to confirm the outbreak group identity. Twenty-three out of 219 strains were classified as PCR false positive by PFGE and SNP analysis of WGS data. Of those 23 strains, 15 strains showed different profiles respect to the outbreak pulsotype (*Apa*I.0246, *Asc*I.0356). The leftover eight strains, despite having the same outbreak pulsotype, were epidemiologically unrelated and showed a number of SNPs above the cut-off (Figure 3).

4. Discussion

In Italy, in 2018, notification rate of invasive listeriosis was 0.29 cases per 100,000 population, lower than the average of European countries (0.47). Indeed, despite *Listeria* rarely exceeding the EU food safety limit tested in ready-to-eat food, many European countries reported rates higher than 0.80 (Estonia, Finland, Spain, Sweden, Denmark, Luxemburg, and Germany) [3].

Rapid identification of contaminated food and food processing industries is critical to stop the diffusion of pathogenic strains and minimize the number of cases in a foodborne outbreak. Depending on the setting (local, national, or international), one or more molecular methods needs to be carried out for typing and subtyping the involved strains.

Between January 2015 and February 2016, a total number of 63 *L. monocytogenes* strains isolated from food samples potentially related to the outbreak were sent to the It NRL Lm in order to identify the source of infection. In February 2016, strains isolated from a sample of hog head cheese (a typical pork-derived meat jelly-seasoned product) were identified as serotype 1/2a and showed 100% genetic similarity with the outbreak pulsotype by PFGE [4]. After that, deep tracing back and forward investigations were carried out and 1348 strains were isolated from both contaminated food and food processing environmental samples.

Before the use of the real-time PCR method, isolates were screened by molecular serogrouping. As described in several studies [22–24], IIa is the main serogroup within strains isolated from food and food processing environments. Accordingly, in our study almost half of the typed strains were IIa flaA+, thus molecular serogrouping was not suitable for the screening purposes. Moreover, considering the number of strains received at the IT NRL Lm and the need to establish the inclusion or exclusion of food or food processing industries in the outbreak, the usual workflow was no more applicable, because the time needed to obtain results of molecular serogrouping coupled with PFGE analysis became too long.

A PCR-based method was developed to decrease the turnaround time for the screening. The real-time PCR method was able to quickly analyze up to 96 strains in 2–4 h and to identify all the strains potentially related to the outbreak. Positive isolates were then further investigated by PFGE and SNP analysis.

Positive results were also observed in 10.5% of strains not related to the outbreak. However, false positive outcomes are generally accepted for the inclusivity scope of a screening assay, designed to limit the number of the strains addressed for further analysis, but absolutely able to include all the strains potentially related to the outbreak.

Applying the described real-time PCR assay, we were able to drastically decrease time and costs of analysis during the outbreak. The developed strategy selected only 219 out of 587 strains, classified as

IIa by molecular serogrouping, which were submitted to SNP analysis and PFGE with priority. As the estimated cost of PFGE per single strain is between 20.9 and 23.3 Euro, depending on the gel size [25], the gain ranged from 7691 and 8574 Euro. Moreover, the rapid response time of the analyses limited economic loss for the food business operators and farms involved. Overall, serogroup/PFGE screening was demonstrated to be less effective than the assay developed in this study.

5. Conclusions

WGS data were used as "starting point" to develop a PCR-based diagnostic test. We demonstrated that our novel real-time PCR assay can screen thousands of *L. monocytogenes* strains, significantly reducing time and costs.

Despite the fact that the method was specifically designed for *L. monocytogenes* strains involved in the listeriosis outbreak that occurred in Central Italy, this case report represents a proof of concept suggesting to the scientific community a new approach for any outbreak management.

WGS data were finally used as "closing point" of the workflow, the SNP-based clustering method for *L. monocytogenes* isolates enabled discrimination of strains indistinguishable in PFGE, but not correlated to the outbreak, and identification of the strains isolated from a hog head cheese sample produced by a facility located in Central Italy as the possible source of human infection.

Author Contributions: Authors contributed equally to this work. All authors have read and agreed to the published version of the manuscript.

Funding: This research was funded by the Italian Ministry of Health.

Acknowledgments: We thank the personnel of the local, regional and central competent authority.

Conflicts of Interest: The authors declare no conflict of interest.

References

1. Radoshevich, L.; Cossart, P. *Listeria monocytogenes*: Towards a complete picture of its physiology and pathogenesis. *Nat. Rev. Microbiol.* **2018**, *16*, 32–46. [CrossRef]
2. Chlebicz, A.; Śliżewska, K. Campylobacteriosis, Salmonellosis, Yersiniosis, and Listeriosis as Zoonotic Foodborne Diseases: A Review. *Int. J. Environ. Res. Public Health* **2018**, *15*, 863. [CrossRef]
3. European Food Safety Authority (EFSA) and European Centre for Disease Prevention and Control. (ECDC). The European Union summary report on trends and sources of zoonoses, zoonotic agents and food-borne outbreaks in 2018. *EFSA J.* **2019**, *17*, 5926. Available online: https://efsa.onlinelibrary.wiley.com/doi/epdf/10.2903/j.efsa.2019.5926 (accessed on 27 December 2019).
4. Duranti, A.; Sabbatucci, M.; Blasi, G.; Acciari, V.A.; Ancora, M.; Bella, A.; Busani, L.; Centorame, P.; Cammà, C.; Conti, F.; et al. A severe outbreak of listeriosis in central Italy with a rare pulsotype associated with processed pork products. *J. Med. Microbiol.* **2018**, *67*, 1351–1360. [CrossRef] [PubMed]
5. Pietzka, A.; Allerberger, F.; Murer, A.; Lennkh, A.; Stöger, A.; Cabal Rosel, A.; Huhulescu, S.; Maritschnik, S.; Springer, B.; Lepuschitz, S.; et al. Whole Genome Sequencing Based Surveillance of *L. monocytogenes* for Early Detection and Investigations of Listeriosis Outbreaks. *Front. Public Health* **2019**, *7*, 139. [CrossRef] [PubMed]
6. Datta, A.R.; Burall, L.S. Serotype to genotype: The changing landscape of listeriosis outbreak investigations. *Food Microb.* **2018**, *75*, 18–27. [CrossRef]
7. Struelens, M.J.; Brisse, S. From molecular to genomic epidemiology: Transforming surveillance and control of infectious diseases. *Eurosurveillance* **2013**, *18*, 20386. [CrossRef]
8. Quainoo, S.; Coolen, J.P.M.; van Hijum, S.A.F.T.; Huynen, M.A.; Melchers, W.J.G.; van Schaik, W.; Wertheim, H.F.L. Whole-Genome Sequencing of Bacterial Pathogens: The Future of Nosocomial Outbreak Analysis. *Clin. Microbiol. Rev.* **2017**, *30*, 1015–1063. [CrossRef]
9. Doumith, M.; Buchrieser, C.; Glaser, P.; Jacquet, C.; Martin, P. Differentiation of the Major *Listeria monocytogenes* Serovars by Multiplex PCR. *J. Clin. Microbiol.* **2004**, *42*, 3819–3822. [CrossRef]
10. Kérouanton, A.; Marault, M.; Petit, L.; Grout, J.; Dao, T.T.; Brisabois, A. Evaluation of a multiplex PCR assay as an alternative method for *Listeria monocytogenes* serotyping. *J. Microbiol. Methods* **2010**, *80*, 134–137. [CrossRef]

11. PulseNet-International. Standard Operating Procedure for Pulsenet PFGE of Listeria monocytogenes. 2013. Available online: http://www.cdc.gov/pulsenet/PDF/listeria-pfge-protocol-508c.pdf (accessed on 21 November 2019).

12. Acciari, V.A.; Torresi, M.; Iannetti, L.; Scattolini, S.; Pomilio, F.; Decastelli, L.; Colmegna, S.; Muliari, R.; Bossù, T.; Proroga, Y.; et al. *Listeria monocytogenes* in smoked salmon and other smoked fish at retail in Italy: Frequency of contamination and strain characterization in products from different manufacturers. *J. Food Prot.* **2017**, *80*, 271–278. [CrossRef] [PubMed]

13. Bolger, A.M.; Lohse, M.; Usadel, B. Trimmomatic: A flexible trimmer for Illumina Sequence Data. *Bioinformatics* **2014**, *30*, 2114–2120. [CrossRef] [PubMed]

14. Bankevich, A.; Nurk, S.; Antipov, D.; Gurevich, A.A.; Dvorkin, M.; Kulikov, A.S.; Lesin, V.M.; Nikolenko, S.I.; Pham, S.; Prjibelski, A.D.; et al. SPAdes: A new genome assembly algorithm and its applications to single-cell sequencing. *J. Comput. Biol.* **2012**, *19*, 455–477. [CrossRef] [PubMed]

15. Seemann, T. Prokka: Rapid prokaryotic genome annotation. *Bioinformatics* **2014**, *30*, 2068–2069. [CrossRef]

16. Page, A.J.; Cummins, C.A.; Hunt, M.; Wong, V.K.; Reuter, S.; Holden, M.T.G.; Fookes, M.; Falush, D.L.; Keane, J.A.; Parkhill, J. Roary: Rapid large-scale prokaryote pan genome analysis. *Bioinformatics* **2015**, *31*, 3691–3693. [CrossRef]

17. Gardner, S.N.; Slezak, T.; Hall, B.G. kSNP3.0: SNP detection and phylogenetic analysis of genomes without genome alignment or reference genomes. *Bioinformatics* **2015**, *31*, 2877–2878. [CrossRef]

18. Morganti, M.; Scaltriti, E.; Cozzolino, P.; Bolzoni, L.; Casadei, G.; Pierantoni, M.; Foni, E.; Pongolini, S. Processing-Dependent and Clonal Contamination Patterns of *Listeria monocytogenes* in the Cured Ham Food Chain Revealed by Genetic Analysis. *Appl. Environ. Microbiol.* **2015**, *82*, 822–831. [CrossRef]

19. Tamura, K.; Stecher, G.; Peterson, D.; Filipski, A.; Kumar, S. MEGA6: Molecular Evolutionary Genetics Analysis version 6.0. *Mol. Biol. Evol.* **2013**, 2725–2729. [CrossRef]

20. Larkin, M.A.; Blackshields, G.; Brown, N.P.; Chenna, R.; McGettigan, P.A.; McWilliam, H.; Valentin, F.; Wallace, I.M.; Wilm, A.; Lopez, R.; et al. Clustal W and Clustal X version 2.0. *Bioinformatics* **2007**, *23*, 2947–2948. [CrossRef]

21. Vaerman, J.L.; Saussoy, P.; Ingargiola, I. Evaluation of Real Time PCR data. *J. Biol. Regul. Homeost. Agents* **2004**, *18*, 212–214.

22. Lee, S.; Chen, Y.; Gorski, L.; Ward, T.J.; Osborne, J.; Kathariou, S. *Listeria monocytogenes* Source Distribution Analysis Indicates Regional Heterogeneity and Ecological Niche Preference among Serotype 4b Clones. *mBio* **2018**, *9*. [CrossRef] [PubMed]

23. Orsi, R.H.; den Bakker, H.C.; Wiedmann, M. *Listeria monocytogenes* lineages: Genomics, evolution, ecology, and phenotypic characteristics. *Int. J. Med. Microbiol.* **2011**, *301*, 79–96. [CrossRef] [PubMed]

24. Pasquali, F.; Palma, F.; Guillier, L.; Lucchi, A.; De Cesare, A.; Manfreda, G. Listeria monocytogenes Sequence Types 121 and 14 Repeatedly Isolated within One Year of Sampling in a Rabbit Meat Processing Plant: Persistence and Ecophysiology. *Front. Microbiol.* **2018**, *9*, 596. [CrossRef]

25. Michelon, D.; Félix, B.; Vingadassalon, N.; Mariet, J.-F.; Larsson, J.T.; Møller-Nielsen, E.; Roussel, S. PFGE Standard Operating Procedures for *Listeria monocytogenes*: Harmonizing the Typing of Food and Clinical Strains in Europe. *Foodborne Pathog. Dis.* **2015**, *12*, 244–252. [CrossRef] [PubMed]

Article

The Influence of Different Smoking Procedures on the Content of 16 PAHs in Traditional Dry Cured Smoked Meat "Hercegovačka Pečenica"

Leona Puljić [1], Krešimir Mastanjević [2,*], Brankica Kartalović [3], Dragan Kovačević [2], Jelena Vranešević [3] and Kristina Mastanjević [2]

[1] The Faculty of Agriculture and Food Technology (APTF) of the University of Mostar, Biskupa Čule bb, 88000 Mostar, Bosnia and Herzegovina; leonapuljic224@gmail.com
[2] Faculty of Food Technology Osijek, Josip Juraj Strossmayer University of Osijek, F. Kuhača 20, 31000 Osijek, Croatia; dragan.kovacevic@ptfos.hr (D.K.); kristinahabschied@gmail.com (K.M.)
[3] Scientific Veterinary Institute Novi Sad, Rumenački put 20, 21000 Novi Sad, Serbia; brankica@niv.ns.ac.rs (B.K.); vet.bmv@hotmail.com (J.V.)
* Correspondence: kmastanj@gmail.com; Tel.: +385-31-224-300

Received: 15 November 2019; Accepted: 11 December 2019; Published: 17 December 2019

Abstract: During smoking, meat products may get contaminated by polycyclic aromatic hydrocarbons (PAH), especially the ones that are smoked in traditional (uncontrolled) conditions. This study aims to evaluate the difference in PAH content in samples of traditional dry cured pork meat products, "Hercegovačka pečenica", produced in (1) a traditional smokehouse and (2) in industrial chambers. The study revealed that the content of the four priority PAHs (PAH4) in samples produced in a traditional smoking manner highly exceeded (up to 10 times) the maximal limits set for PAHs (12 μg/kg). PAH4 in all samples subjected to industrial smoking procedures was below the limit of quantification. All samples had below-the-limit-of-quantification values for Benzo[a]pyrene. The surface layer of the samples produced in traditional conditions had the highest total content of PAH16. The inner parts of all samples, whether traditional or industrial, had significantly lower PAH16 concentration than the surface layer.

Keywords: PAH content; Hercegovačka pečenica; traditional smoking; industrial smoking

---

1. Introduction

Hercegovačka pečenica is a traditional dry cured pork meat product produced in Herzegovina (a southern region of Bosnia and Herzegovina) in a specific way. Smoked pork loin is a cured meat product obtained from a long back muscle lat, musculus longissimus dorsi, without bones and skin, using a salting or brining process. Other spices or herbs may be added optionally. The smoking, drying, and ripening of the product takes from one to three months. Specific geographical conditions of Herzegovina (the influence of the sub-Mediterranean climate in particular) with a significant influence of the Mediterranean and in some parts of the continental climate, distinguishes Hercegovačka pečenica from similar products originating from other areas. The characteristic of this microclimate is bora and sirocco winds, which meet and collide only in this area. The peculiarity of this product is the smoke of hornbeam and beech trees from the high altitudes of the nearby mountains. Meat smoking is widely used in the production of traditional dry cured meat products from Herzegovina. One of these is the traditional Hercegovačka pečenica. A typical diet in Herzegovina involves different kinds of smoked meat products. The direct smoking during the drying process is still the most commonly used technique with traditional producers.

The production of Hercegovačka pečenica mainly takes place in smokehouses on family farms. Traditional smoking involves treating of pre-salted pork loins with wood smoke. During smoking, particular sensorial features in terms of taste, color, and aroma are formed (phenol derivatives, carbonyls, organic acids, among others) [1–3]. In addition, smoking improves conservation due to its dehydrating, bactericidal, and antioxidant properties [2]. Incomplete wood combustion during the process of smoking is responsible for the production of significant amounts of polycyclic aromatic hydrocarbons (PAHs), a potential health hazard [4,5]. They have two or more fused aromatic rings and are known as cancer-causing agents [6]. PAHs differ in their carcinogenicity. Even though some of them are regarded as non-carcinogenic, they may increase the carcinogenicity of other PAHs [7]. PAHs are highly lipophilic, and thus can be found in meat products subjected to smoking [8–10]. The factors that limit the occurrence of PAHs during the smoking process are various. The most important ones are the following: the technique of smoking, choice of wood, the duration of smoke exposure, and the type of food itself [11–13]. Codex alimentarius [14] guidelines for optimal smoking should be applied to reduce the PAH concentrations in processed foods.

The European food safety authority (EFSA) decided that the concentrations of benzo[a]pyrene (BaP) and the sum of the concentrations of four PAHs: benzo[a]pyrene (BaP), benz[a]anthracene (BaA), benzo[b]fluoranthene (BbF), and chrysene (Chry) (PAH4) [15], will be considered a reference for determination of PAHs in food. According to Bosnian and Herzegovinian regulations on maximum levels for certain contaminants in food [16], complied with the European commission (EU) regulation no. 835/2011 [17], the maximum permissible concentration of BaP in meat products is set at 2 µg/kg and the sum of PAH4 concentrations should not exceed 12 µg/kg. Although there are studies across Europe and some developing countries concerning the carcinogenic potential and occurrence of PAHs in food products [1,5,10,18–22], there is a lack of information about PAH occurrence in meat products and smoked meat products, in particular in Bosnia and Herzegovina. Thus, the aim of this study was to investigate and compare the differences in PAH content in the samples of "Hercegovačka pečenica" subjected to traditional and industrial smoking processes.

2. Materials and Methods

2.1. Preparation of Smoked Meat Samples

Raw material processing is performed using traditional technology. The samples were made from pork long back muscle lat, musculus longissimus dorsi. Immediately before salting, the weight of each individual raw loin was determined by weighing (approximately 3 kg), followed by salting with a 50:50 mixture of rock and nitrite salt. The meat was salted once manually by adding an indefinite amount of salt over it. The meat was then left in the salt for seven days in a cooling chamber at a temperature of 4 °C. After the salting processes were completed, the loins were washed in water and transferred to a drying and smoking room where they were drained and tempered for the next 12 to 20 h. The description of smoking conditions is shown in Table 1.

Table 1. Production conditions and variables of Hercegovačka pečenica samples.

Abbreviation	Batch	Smoking Time (days)	Sampling Position
TS-SP	Traditional smoking at 3 m distance from fire	20	surface
TS-MP	Traditional smoking at 3 m distance from fire	20	inner
IS-SP	Industrial smoking	3	surface
IS-MP	Industrial smoking	3	inner

When the traditional smoking method was applied, raw and pre-salted pork loins (three replicates) were put at three meters distance from open fire. The smoking was carried out by combustion of dry hard wood (beech) every day for the first six days (for 6–8 h), and every two or three days (for 2–3 h) for the next fourteen days. It lasted for 20 days in ambient conditions. The temperature ranged from 3.5 to 11.2 °C (average 6.9 °C) and relative humidity from 61.3 to 90.5% (average 74.2%).

In industrial production, pre-salted pork loins (three replicates), were kept in a ripozo chamber for 13 days at the average temperature of 1 °C and relative humidity of 60% until they lost about 15% of the initial weight. A smoking chamber (Mauting, Czech Republic) was used for smoking in industrial production. The chamber was equipped with a heating plate smoke generator where beech sawdust was placed. During industrial smoking, the average temperature was 19.0 °C and average relative humidity was 74.4%. The pork loins were smoked for 4 h a day (8 × 30 min) for three days.

After the smoking, pork loins were placed in a ripening chamber (Mauting, Valtice, Czech Republic). Drying and ripening was set at the average temperature of 15.0 °C and relative humidity was 74.2%. The sampling was carried out at the end of the smoking process (on the 20th day for when traditional procedure was used in and on the 3rd day in the case of the industrial). The total production time for both production procedures (industrial and traditional) was 45 days. All sampling was performed in triplicate. Before the PAH determination, the samples were packed in glass jars and stored in the dark at −30 °C until the analysis was performed, approximately 1 week after the sampling. All the analyses were done in triplicate.

2.2. GC-MS Analysis

Sample preparation for GC_MS analysis and chromatographic separation of 16 PAH (Nap—naphthalene; Anl—acenaphthylene; Ane—acenaphtene; Flu—fluorene; Ant—anthracene; Phen—phenanthrene; Flt—fluoranthene; BaA—benzo[a]anthracene; Pyr—pyrene; Chry—chrysene; BbF—benzo[b]fluoranthene; BkF—benzo[k]fluoranthene; BaP—benzo[a]pyrene; DahA—dibenzo[a;h] anthracene; BghiP—benzo[g;h;i]-perylene; InP—indeno[1;2;3-cd]pyrene) were conducted according to Mastanjević et al. [10]. The average values for precision, reproducibility, accuracy, linearity, LOQ, and LOD for PAH method validation can be found in Supplementary Table S1.

2.3. Statistical Analysis

Analysis of variance (ANOVA) and Fisher's least significant difference (LSD), with significance defined at $p < 0.05$, were performed for all measured data. Statistica 12.7 (StatSoft Inc. Tulsa, OK, USA, 2015) was used for statistical analysis.

3. Results and Discussion

The raw material (pork loins) used for production of Hercegovačka pečenica contained only light PAHs (Nap, Anl, Ane, Flu, and Ant). The other eleven PAHs were below the limit of quantification. Ant was the PAH with the highest concentration of 21.6 µg/kg in raw meat. Other PAHs determined in raw pork loin samples ranged as follows: Nap 10.87 µg/kg, Anl 5.88 µg/kg, Ane 20.0 µg/kg, and Flu 5.46 µg/kg. In Slavonska kobasica, Mastanjević et al. [10] reported similar values for PAHs in raw materials. On the other hand, Djinovic et al. [23] reported lower values of PAHs in raw meat used for the production of different types of smoked meat products from Serbia. According to Ciganek and Neca (2006) [24], the PAHs found in animal tissues are a result of environmental contamination, thus the PAHs in raw pork meat in this research can be attributed to the contamination of feed used in pig breeding.

Mean values of PAHs determined in dry cured loins at the end of the smoking period (the 3rd and 20th day of production) are shown in Table 2. For all batches, PAHs contamination levels are presented for external/surface and inner parts.

PAH content determined in loins (external and inner parts) at the end of the smoking period in the samples subjected to traditional production (20 days of production, open fire) involved Nap, Anl, Flu, Ant, Phen, Flt, BaA, Pyr, BbF, BkF, and BghiP. On the other hand, PAHs determination in loins at the end of the smoking period in the samples subjected to industrial production (3 day production) resulted in NaP and Anl. The other analyzed PAHs were below the limit of quantification for all the sample groups. These results are in accordance with the previous research reports regarding smoked meat products [23,25], and different smoked sausages [2,10,26–30]. The most abundant light PAH in all samples appeared to be Phen, which ranged from 1029 µg/kg for batch TS-SP to <LOQ for batch IS-SP, showing a difference with statistical significance ($p < 0.05$) between batch TS-SP and all other sample groups. Groups TS-MP, IS-MP, and IS-SP did not show a difference with statistical significance for Phen content. Nap concentrations were between 26.7 µg/kg (TS-MP) and 35.5 µg/kg (IS-SP) and no difference with statistical significance was found between groups.

The content of Anl was between 21.2 µg/kg (TS-MP) and 590 µg/kg (TS-SP), showing a difference with statistical significance ($p < 0.05$) between TS-SP and all other sample groups. Flu concentrations ranged from below LOQ (IS-SP) to 406 µg/kg, (TS-SP), showing a difference with statistical significance ($p < 0.05$) between TS-SP and all other sample groups. Ant content ranged from below LOQ (IS-SP) to 242 µg/kg (TS-SP), showing a difference with statistical significance ($p < 0.05$) between TS-SP and all other sample groups. Flt concentrations ranged from below LOQ (IS-SP) to 82.7 (TS-SP) µg/kg with a difference with statistical significance ($p < 0.05$) detected between TS-SP and all other sample groups.

Table 2. 16 PAHs (µg/kg) in Hercegovačka pečenica at the end of smoking process.

PAH	TS-SP	TS-MP	IS-SP	IS-MP
Nap	28.1 [a]	26.7 [a]	35.5 [a]	30.9 [a]
Anl	590 [a]	21.2 [b]	31.5 [b]	26.9 [b]
Ane	-*	-*	-*	-*
Flu	406 [a]	16.0 [b]	-*	-*
Ant	242 [a]	8.71 [b]	-*	-*
Phen	1029 [a]	44.8 [b]	-*	-*
Flt	82.7 [a]	5.44 [b]	-*	-*
BaA	30.9 [a]	11.4 [b]	-*	-*
Pyr	54.5 [a]	2.48 [b]	-*	-*
Chry	-*	-*	-*	-*
BbF	1.5 [a]	1.29 [a]	-*	-*
BkF	4.09 [b]	4.89 [a]	-*	-*
BaP	-*	-*	-*	-*
DahA	-*	-*	-*	-*
BghiP	2.77 [b]	3.00 [a]	-*	-*
Inp	-*	-*	-*	-*
∑PAH4	32.5 [a]	12.7 [b]	-*	-*
∑PAH16	2474 [a]	145 [b]	67.0 [c]	57.9 [d]

PAHs: polycyclic aromatic hydrocarbons; TS-SP: Traditional smoking at 3 m distance from fire (surface); TS-MP: Traditional smoking at 3 m distance from fire (inner); IS-SP: Industrial smoking (surface); IS-MP: Industrial smoking (inner). Means within rows with different superscripts are significantly different ($p < 0.05$); -*: < LOQ (limit of quantification); ∑PAH4: BaA (benzo[a]anthracene); Chry (chrysene); BbF (benzo[b]fluoranthene), and BaP (benzo[a]pyrene) ∑PAH16: Nap (naphthalene), Anl (acenaphthylene), Ane (acenaphtene), Flu (fluorene), Ant (anthracene), Phen (phenanthrene), Flt (fluoranthene), BaA, Pyr (pyrene), Chry, BbF, BkF (benzo[k]fluoranthene), BaP, DahA (dibenzo[a;h]anthracene), BghiP (benzo[g;h;i]-perylene), and InP (indeno[1;2;3-cd]pyrene).

The results for BaA were between below LOQ (IS-SP) and 30.9 µg/kg (TS-SP) with a difference of statistical significance ($p < 0.05$) between all groups except between IS-MP and IS-SP. The content of Pyr was quantified below LOQ (IS-SP) to the maximum value of 54.5 µg/kg in sample TS-MP, with a significant difference ($p < 0.05$) between TS-SP and all other sample groups. BbF concentrations were in the range from below LOQ (IS-SP) to 1.54 µg/kg, (TS-SP) with the difference with statistical significance ($p < 0.05$) between traditional (TS-SP, TS-MP) and industrial (IS-SP, IS-MP) meat products.

The content of BkF was between below LOQ for IS-SP and 4.89 µg/kg in TS-MP. A significant statistical difference ($p < 0.05$) was noted between all groups except between IS-MP and IS-SP. BghiP concentrations were in the range from below LOQ (IS-SP) to 3.00 µg/kg, (TS-MP) and a significant statistical difference ($p < 0.05$) was detected between all groups except between IS-MP and IS-SP.

The sum of PAH16 ranged from 57.9 µg/kg for samples smoked in industrial conditions to 2474 µg/kg for samples smoked in traditional conditions. Mastanjević et al. [10] reported lower summation values of PAH16 for the samples at the end of the smoking phase, smoked in traditional conditions (509 µg/kg), but a higher sum of PAH16 (114 µg/kg) for samples subjected to industrial conditions. On the other hand, Djinovic et al. [23], reported lower PAH16 concentration in different smoked meat products from Serbia, measured at the end of the smoking process. PAH4 content ranged as follows: BaA < LOQ–30.9 µg/kg, Chry < LOQ, BbF < LOQ–1.54 µg/kg, BaP < LOQ, PAH4 < LOQ–32.5 µg/kg. In all sample groups, BaP content was lower than 2 µg/kg. The TS-MP and TS-SP samples had higher concentrations than the prescribed PAH4 content (12.7 µg/kg and 32.5 µg/kg).

The content of the four priority PAHs in all samples produced using traditional smoking techniques exceeded maximal limits set by Bosnian and Herzegovinian Regulation on maximum levels for certain contaminants in food [16], complied with the EC Regulation No. 835/2011 [17]. Such high amounts of PAH have rarely been reported before. However, it corresponds to the values reported for smoking under uncontrolled technological conditions, typical for households and developing countries [6,30]. On the other hand, the amount of the four priority PAHs in all the samples produced using industrial smoking procedures were < LOQ.

Sixteen PAHs in finished dry cured loins are presented in Table 3. The same PAH content (Nap, Anl, Flu, Ant, Phen, Flt, BaA, Pyr, BbF, BkF, and Bghip) was determined in traditional smoked "Hercegovačka pečenica" at the end of the production process. Contrary to this, in industrial smoked samples, only Nap and Anl were detected. The most abundant PAH in traditional smoked samples was Ant, where the content ranged from 2925 µg/kg in the surface of the product to 60.3 µg/kg in the inner part with difference with statistical significance of ($p < 0.05$) between groups. In industrially smoked samples, the most abundant PAH was Anl and it ranged from 3.80 µg/kg in the inner part of the product to 19.3 µg/kg on the surface and no difference with statistical significance ($p > 0.05$) between groups was detected. Nap concentrations were between 3.77 µg/kg (IS-MP) and 718 µg/kg (TS-SP), showing a difference with statistical significance ($p < 0.05$) between TS-SP and all other sample groups. The content of Anl was between 3.80 µg/kg (IS-MP) and 2515 µg/kg (TS-SP). The difference with statistical significance ($p < 0.05$) was detected between TS-SP and all other sample groups. Flu concentrations were in the range from <LOQ (IS-SP) to 701 µg/kg, (TS-SP) with the difference of statistical significance ($p < 0.05$) between TS-SP and all other sample groups. Phen content ranged from <LOQ (IS-SP) to 807 (TS-SP) µg/kg with the difference with statistical significance ($p < 0.05$) between TS-SP and all other sample groups. Flt concentrations ranged from <LOQ (IS-SP) to 237 µg/kg (TS-SP) with the difference with statistical significance ($p < 0.05$) between TS-SP and all other sample groups. BaA concentrations were between below LOQ (IS-SP) and 123 µg/kg (TS-SP) with difference with statistical significance ($p < 0.05$) between TS-SP and all other sample groups. The content of Pyr was between <LOQ (IS-SP) and 187 µg/kg (TS-SP) with difference with statistical significance ($p < 0.05$) between TS-SP and all other sample groups. BbF concentrations were in the range from <LOQ (IS-SP) to 2.36 µg/kg (TS-MP) showing a difference with statistical significance between traditional (TS-SP, TS-MP) and industrial (IS-SP, IS-MP) meat groups. The content of BkF was between <LOQ (IS-SP) and 6.43 µg/kg (TS-MP) with difference with statistical significance ($p < 0.05$) between all groups except between groups IS-MP and IS-SP. BghiP concentrations were in the range from <LOQ (IS-SP) to 2.79 µg/kg, (TS-MP) with difference with statistical significance ($p < 0.05$) between all groups except between the groups IS-MP and IS-SP.

Table 3. 16 PAHs (µg/kg) in the finished product Hercegovačka pečenica.

PAH	TS-SP	TS-MP	IS-SP	IS-MP
Nap	718 [a]	54.8 [b]	6.33 [b]	3.77 [b]
Anl	2515 [a]	30.2 [b]	19.3 [b]	3.80 [b]
Ane	-*	-*	-*	-*
Flu	701 [a]	13.8 [b]	-*	-*
Ant	2925 [a]	60.3 [b]	-*	-*
Phen	807 [a]	11.6 [b]	-*	-*
Flt	237 [a]	5.79 [b]	-*	-*
BaA	123 [a]	5.41 [b]	-*	-*
Pyr	187 [a]	4.10 [b]	-*	-*
Chry	-*	-*	-*	-*
BbF	2.25 [a]	2.36 [a]	-*	-*
BkF	4.38 [b]	6.43 [a]	-*	-*
BaP	-*	-*	-*	-*
DahA	-*	-*	-*	-*
BghiP	2.60 [b]	2.79 [a]	-*	-*
Inp	-*	-*	-*	-*
∑PAH4	125a	7.77b	-*	-*
∑PAH16	8225a	197b	25.6c	7.6d

TS-SP—Traditional smoking at 3 m distance from fire (surface); TS- MP—Traditional smoking at 3 m distance from fire (inner); IS-SP—Industrial smoking (surface); IS-MP—Industrial smoking (inner). Means within rows with different superscripts are significantly different (*p* < 0.05); -*: < LOQ (limit of quantification); ∑PAH4: BaA; Chry; BbF, and BaP ∑PAH16: Nap, Anl, Ane, Flu, Ant, Phen, Flt, BaA, Pyr, Chry, BbF, BkF, BaP, DahA, BghiP, and InP.

External/surface parts of Hercegovačka pečenica smoked industrially and traditionally showed ∑ 16 PAHs 25.6 µg/kg and 8225 µg/kg. The inner parts of all samples (both industrial and traditional) had a significantly lower total PAHs contamination levels (197 µg/kg for traditionally smoked samples and 7.57 µg/kg for industrial method of smoking). Ciecierska et al. (2007) [25] reported much lower concentrations of PAH 15 in both traditionally and industrially smoked raw cured loins (external part 10.7 µg/kg and inner part 1.52 µg/kg for traditional production and 9.42 µg/kg and 2.59 µg/kg for industrial production). Higher values for PAH content in Hercegovačka pečenica are probably a result of intensely extended smoking.

In general, the concentrations (in both external and inner parts) of all determined light PAHs at the end of the production process of Hercegovačka pečenica were higher than concentrations determined at the end of smoking. Mastanjević et al. [10] reported similar results for Slavonska kobasica. They presumed that this is due to the dehydration. At the end of the industrial production, the concentrations of all determined PAHs were lower than concentrations determined at the end of smoking.

The amount of the four priority PAHs in the samples produced by traditional smoking highly exceeded maximal limits set by Regulation (EU) No. 835/2011 (12 µg/kg) up to 10 times. In all samples produced by industrial smoking, the PAH4 content was below the limit of quantification. BaP concentration in investigated samples was below the limit of quantification. The highest total content of PAH16 (8225 µg/kg) was determined on the surface samples produced in traditional smokehouses at the end of the production. According to the results, one of the main factors which contributed to high levels of PAH in Hercegovačka pečenica is the smoking technique. Another important factor is smoking duration. The longer the samples are smoked, the higher concentration of the PAH can be expected [31,32]. In this research, the samples smoked in a traditional manner (open fire, 20 days of smoking) had significantly higher levels of 16 PAH content than industrially smoked samples. Also, the wood type used for smoking can significantly affect the PAH content in smoked meat products [2].

4. Conclusions

The use of traditional smoking method resulted in higher PAH contamination than the industrial. At the end of production, the inner parts of all smoked samples produced using both methods retained significantly lower total PAHs concentration, as well as less individual PAHs than the surface layer. The amount of the four priority PAHs in samples subjected to traditional smoking highly exceeded maximum limits set by the Regulation (EU) No 835/2011 (12 µg/kg) by up to 10 times. The consumption of this kind of products can be potentially harmful to human health and that is the reason why the ALARA (as low as reasonably achievable) principle is in force in the EU [33]. On the other hand, the amounts of the four priority PAHs in all samples subjected to industrial smoking processes were below the limit of quantification. The result of this study indicated that, in order to decrease the level of PAHs and reduce the risk of PAHs occurrence in smoked meat products, local producers should learn how to use the improved/novel smoking techniques and adjust the smoking parameters. This should result in safer smoked meat products.

Supplementary Materials: The following are available online at http://www.mdpi.com/2304-8158/8/12/690/s1, Table S1: The average values for precision, reproducibility, accuracy, linearity, LOQ and LOD for PAH method validation.

Author Contributions: Conceptualization, K.M. (Krešimir Mastanjević); methodology, B.K.; software, K.M. (Krešimir Mastanjević); validation, B.K., J.V.; investigation, L.P.; data curation, D.K.; writing—original draft preparation, L.P.; writing—review and editing, K.M. (Kristina Mastanjević).; supervision, K.M. (Krešimir Mastanjević).

Funding: This research received no external funding.

Conflicts of Interest: The authors declare no conflict of interest.

References

1. Roseiro, L.C.; Gomes, A.; Santos, C. Influence of processing in the prevalence of polycyclic aromatic hydrocarbons in a Portuguese traditional meat product. *Food Chem. Toxicol.* **2011**, *49*, 1340–1345. [CrossRef] [PubMed]
2. Škaljac, S.; Jokanović, M.; Tomović, V.; Ivić, M.; Tasić, T.; Ikonić, P.; Šojić, B.; Džinić, N.; Petrović, L. Influence of smoking in traditional and industrial conditions on colour and content of polycyclic aromatic hydrocarbons in dry fermented sausage "Petrovská klobása". *LWT* **2018**, *87*, 158–162. [CrossRef]
3. Bogdanović, T.; Pleadin, J.; Petričević, S.; Listeš, E.; Sokolić, D.; Marković, K.; Ozogul, F.; Šimat, V. The occurrence of polycyclic aromatic hydrocarbons in fish and meat products of Croatia and dietary exposure. *J. Food Compos. Anal.* **2019**, *75*, 49–60. [CrossRef]
4. Alomirah, H.; Al-Zenki, S.; Al-Hooti, S.; Zaghloul, S.; Sawaya, W.; Ahmed, N.; Kannan, K. Concentrations and dietary exposure to polycyclic aromatic hydrocarbons (PAHs) from grilled and smoked foods. *Food Control* **2011**, *22*, 2028–2035. [CrossRef]
5. Ledesma, E.; Rendueles, M.; Díaz, M. Contamination of meat products during smoking by polycyclic aromatic hydrocarbons: Processes and prevention. *Food Control* **2016**, *60*, 64–87. [CrossRef]
6. Šimko, P. Determination of polycyclic aromatic hydrocarbons in smoked meat products and smoke flavouring food additives. *J. Chromatogr. B* **2002**, *770*, 3–18. [CrossRef]
7. Hwang, K.; Woo, S.; Choi, J.; Kim, M. Survey of polycyclic aromatic hydrocarbons in marine products in Korea using GC/MS. *Food Addit. Contam.* **2012**, *5*, 1–7. [CrossRef]
8. Babić, J.M.; Kartalović, B.D.; Škaljac, S.; Vidaković, S.; Ljubojević, D.; Petrović, J.M.; Ćirković, M.A.; Teodorović, V. Reduction of polycyclic aromatic hydrocarbons in common carp meat smoked in traditional conditions. *Food Addit. Contam.* **2018**, *11*, 208–213. [CrossRef]
9. Abdel-Shafy, H.I.; Mansour, M.S.M. A review on polycyclic aromatic hydrocarbons: Source, environmental impact, effect on human health and remediation. *Egypt. J. Pet.* **2016**, *25*, 107–123. [CrossRef]
10. Mastanjević, K.; Kartalović, B.; Petrović, J.; Novakov, N.; Puljić, L.; Kovačević, D.; Jukić, M.; Lukinac, J.; Mastanjević, K. Polycyclic aromatic hydrocarbons in traditional smoked sausage Slavonska kobasica. *J. Food Compos. Anal.* **2019**, *83*, 103282. [CrossRef]

11. Hitzel, A.; Pöhlmann, M.; Schwägele, F.; Speer, K.; Jira, W. Polycyclic aromatic hydrocarbons (PAH) and phenolic substances in meat products smoked with different types of wood and smoking spices. *Food Chem.* **2013**, *139*, 955–962. [CrossRef] [PubMed]

12. Fasano, E.; Yebra-Pimentel, I.; Martínez-Carballo, E.; Simal-Gándara, J. Profiling, distribution and levels of carcinogenic polycyclic aromatic hydrocarbons in traditional smoked plant and animal foods. *Food Control* **2016**, *59*, 581–590. [CrossRef]

13. Malarut, J.; Vangnai, K. Influence of wood types on quality and carcinogenic polycyclic aromatic hydrocarbons (PAHs) of smoked sausages. *Food Control* **2017**, *85*, 98–106. [CrossRef]

14. Codex Alimentarius Commission. Proposed Draft Code of Practice for the Reduction of Contamination of Food with Polycyclic Aromatic Hydrocarbons (PAH) from Smoking and Direct Drying Processes. 2008. Available online: https://www.livsmedelsverket.se/globalassets/produktion-handel-kontroll/lokaler-hanter ing-hygien/codex-alimentarius-forslag-till-praxis-for-att-minska-pah-vid-rokning.pdf?AspxAutoDetectC ookieSupport=1 (accessed on 3 October 2019).

15. European Food Safety Authority. EFSA Scientific opinion of the panel on contaminants in the food chain on a request from the European Commission on polycyclic aromatic hydrocarbons in food. *EFSA J.* **2008**, *724*, 1–114.

16. Regulation No 68/14, 79/16, 84/18. Pravilnik o najvećim dopuštenim količinama određenih kontaminanata u hrani ("Službeni Glasnik BiH", br. 68/14, 79/16, 84/18). Available online: http://www.fsa.gov.ba/fsa/images/pravni-propisi/hr-Pravilnik_o_najve%C4%87im_dopu%C5%A1te nim_koli%C4%8Dinama_odre%C4%91enih_kontaminanata_u_hrani_68-14.pdf (accessed on 3 October 2019).

17. Commission Regulation (EU) No 835/2011. Available online: https://op.europa.eu/en/publication-detail/-/p ublication/6a58ffa2-7404-4acf-b1df-298f611f813d/language-en (accessed on 15 July 2019).

18. Farhadian, A.; Jinap, S.; Abas, F.; Sakar, Z.I. Determination of polycyclic aromatic hydrocarbons in grilled meat. *Food Control* **2010**, *21*, 606–610. [CrossRef]

19. Wretling, S.; Eriksson, A.; Eskhult, G.A.; Larsson, B. Polycyclic aromatic hydrocarbons (PAHs) in Swedish smoked meat and fish. *J. Food Compos. Anal.* **2010**, *23*, 264–272. [CrossRef]

20. Zachara, A.; Gałkowska, D.; Juszczak, L. Contamination of smoked meat and fish products from Polish market with polycyclic aromatic hydrocarbons. *Food Control* **2017**, *80*, 45–51. [CrossRef]

21. Novakov, N.J.; Mihaljev, Ž.A.; Kartalović, B.D.; Blagojević, B.J.; Petrović, J.M.; Ćirković, M.A.; Rogan, D.R. Heavy metals and PAHs in canned fish supplies on the Serbian market. *Food Addit. Contam.* **2017**, *10*, 208–215. [CrossRef]

22. Petrović, J.; Kartalović, B.; Ratajac, R.; Spirić, D.; Djurdjević, B.; Polaček, V.; Pucarević, M. PAHs in different honeys from Serbia. *Food Addit. Contam.* **2019**, *12*, 1–23. [CrossRef]

23. Djinovic, J.; Popovic, A.; Jira, W. Polycyclic aromatic hydrocarbons (PAHs) in different types of smoked meat products from Serbia. *Meat Sci.* **2008**, *80*, 449–456. [CrossRef]

24. Ciganek, M.; Neca, J. Polycyclic aromatic hydrocarbons in porcine and bovine organs and tissues. *Vet. Med.* **2006**, *51*, 239–247. [CrossRef]

25. Ciecierska, M.; Obiedziński, M. Influence of smoking process on polycyclic aromatic hydrocarbons' content in meat products. *Acta Sci. Pol. Technol. Aliment.* **2007**, *6*, 17–28.

26. Lorenzo, J.M.; Purriños, L.; Bermudez, R.; Cobas, N.; Figueiredo, M.; García Fontán, M.C. Polycyclic aromatic hydrocarbons (PAHs) in two Spanish traditional smoked sausage varieties: "Chorizo gallego" and "Chorizo de cebolla". *Meat Sci.* **2011**, *89*, 105–109. [CrossRef] [PubMed]

27. Lorenzo, J.M.; Purriños, L.; Fontán, M.C.G.; Franco, D. Polycyclic aromatic hydrocarbons (PAHs) in two Spanish traditional smoked sausage varieties: "Androlla" and "Botillo". *Meat Sci.* **2010**, *86*, 660–664. [CrossRef] [PubMed]

28. Gomes, A.; Santos, C.; Almeida, J.; Elias, M.; Roseiro, L.C. Effect of fat content, casing type and smoking procedures on PAHs contents of Portuguese traditional dry fermented sausages. *Food Chem. Toxicol.* **2013**, *58*, 369–374. [CrossRef]

29. Škaljac, S.; Petrović, L.; Tasić, T.; Ikonić, P.; Jokanović, M.; Tomović, V.; Džinić, N.; Šojić, B.; Tjapkin, A.; Škrbić, B. Influence of smoking in traditional and industrial conditions on polycyclic aromatic hydrocarbons content in dry fermented sausages (Petrovská klobása) from Serbia. *Food Control* **2014**, *40*, 12–18. [CrossRef]

30. Slámová, T.; Fraňková, A.; Hubáčková, A.; Banout, J. Polycyclic aromatic hydrocarbons in Cambodian smoked fish. *Food Addit. Contam.* **2017**, *10*, 248–255. [CrossRef]

31. Essumang, D.K.; Dodoo, D.K.; Adjei, J.K. Effect of smoke generation sources and smoke curing duration on the levels of polycyclic aromatic hydrocarbon (PAH) in different suites of fish. *Food Chem. Toxicol.* **2013**, *58*, 86–94. [CrossRef]

32. Babić, J.; Vidaković, S.; Bošković, M.; Glišić, M.; Kartalović, B.; Škaljac, S.; Nikolić, A.; Ćirković, M.; Teodorović, V. Content of Polycyclic Aromatic Hydrocarbons in Smoked Common Carp (*Cyprinus Carpio*) in Direct Conditions Using Different Filters vs. Indirect Conditions. *Polycycl. Aromat. Compd.* **2018**. [CrossRef]

33. Commission Regulation (EU) No XX/2019. Available online: https://op.europa.eu/en/publication-detail/-/publication/820239bb-91db-11e9-9369-01aa75ed71a1/language-en/format-HTML/source-108051054 (accessed on 14 August 2019).

Article

Accreditation Procedure for *Trichinella* spp. Detection in Slaughterhouses: The Experience of an Internal Laboratory in Italy

Maria Schirone [1], Pierina Visciano [1,*], Alberto Maria Aldo Olivastri [2], Maria Paola Sgalippa [3] and Antonello Paparella [1]

[1] Faculty of Bioscience and Technology for Food, Agriculture and Environment, University of Teramo, 64100 Teramo, Italy; mschirone@unite.it (M.S.); apaparella@unite.it (A.P.)
[2] Azienda Sanitaria Unica Regionale Marche, 63100 Ascoli Piceno, Italy; albertoolivastri@libero.it
[3] FreeLance Veterinary, 63100 Ascoli Piceno, Italy; mariapaola.sgal@libero.it
*   Correspondence: pvisciano@unite.it; Tel.: +39-086-126-6911

Received: 8 May 2019; Accepted: 4 June 2019; Published: 6 June 2019

Abstract: Trichinellosis is a severe foodborne zoonotic disease due to the consumption of undercooked meat containing *Trichinella* spp. larvae. According to Commission Regulation (EU) No 1375/2015, domestic pigs, farmed wild boar, and horses must be tested for the presence of the parasite in the muscles as part of post-mortem examination. In this study, the accreditation procedure and the maintenance of the certificate for internal laboratory attached to a slaughterhouse are described. The main advantages of such accreditation are represented by the possibility to obtain fast results in order to process carcasses quickly, whereas the difficulties for the technician are linked to performing proficiency testing and following training courses. This program can be considered particularly useful for surveillance and food safety purposes.

Keywords: *Trichinella* spp.; slaughterhouse; accreditation; proficiency testing

1. Introduction

Trichinellosis is one of the most important foodborne parasitic diseases caused by the consumption of raw or undercooked meat of swine, horses, and wild animals infected by the nematode larvae of *Trichinella* spp. [1]. The genus *Trichinella* involves nine species and three genotypes, distinguished in encapsulated (*Trichinella spiralis*, *Trichinella nativa*, *Trichinella britovi*, *Trichinella murrelli*, *Trichinella* T6, *Trichinella nelsoni*, *Trichinella* T8, *Trichinella* T9, and *Trichinella patagoniensis*), transmissible only to mammals, and non-encapsulated (*Trichinella pseudospiralis*, *Trichinella papuae*, and *Trichinella zimbabwensis*), infective to mammals, birds, and reptiles. The first three cited species as well as *T. pseudospiralis* show a high pathogenicity to humans. With regards to their distribution, *T. spiralis* is reported worldwide and can have a domestic and sylvatic life cycle. Also *T. nativa* and *T. britovi* have been described in Europe [2,3]. Turiac et al. [4] reported that *Trichinella* spp. were detected in 354 animals from 1985 to 2016 in Italy, with percentages of 97.5 for *T. britovi*, 2.2 for *T. pseudospiralis*, and 0.3 for *T. spiralis*.

Trichinella spp. recognize a single host in which two generations of the parasite can occur but it does not involve a free-living stage [5,6]. The life cycle consists of three phases: intestinal, migrant, and muscular [7]. In the muscular phase, the larvae can survive from about 1–2 to 10–15 years in hosts because they form a nurse cell–parasite complex into myocytes after de-differentiation and re-differentiation of muscle cells [8,9].

The symptoms of trichinellosis in humans can be different on the basis of the phase of the life cycle of the parasite. The disease recognizes five clinical forms: severe, moderately severe, benign, abortive, and asymptomatic [10]. During the intestinal acute phase, diarrhea and abdominal pain are the most common signs preceding fever and myalgia. Migrating larvae cause immunological and metabolic disturbances, with eosinophilia and release of histamine, serotonin, and prostaglandins [11]. Other symptoms of the acute stage are pyrexia, periorbital or facial oedema, and myalgia. Major complications of trichinellosis can be cardiovascular (i.e., myocarditis and tachycardia), neurological (meningitis or encephalopathy), respiratory (dyspnea, pneumonia, and obstructive bronchitis); death from trichinellosis is rare [10]. In Europe, trichinellosis cases in humans were 224 in 2017, with a notification rate of 0.03 cases per 100,000 populations representing an increase of 50% compared with 2016 [12].

The domestic pigs may ingest the parasite by feeding non-cooked scraps or offal from slaughtering. The transmission to humans is generally linked to the consumption of uncooked or undercooked meat from pigs or hunted wild animals ingesting infected rodents and wildlife. During the period 2004–2014, Badagliacca et al. [13] reported 91 wild mammals resulting positive, namely, wolf, red fox, wild boar, stone marten, pine marten, and wildcat.

The monitoring of the presence of *Trichinella* spp. in animal species susceptible to infection has been implemented since 2006 according to Commission Regulation (EC) No 2075/2005 [14], which allowed risk-based *Trichinella* spp. testing of domestic pigs with derogation for those raised in farms under negligible risk [15]. This Regulation has been repealed by Commission Regulation (EU) No 1375/2015 [16], laying down rules for the sampling of carcasses of domestic swine, horses, wild boar, and other farmed and wild animal species susceptible to *Trichinella* spp. infestation. It also provides for reference and equivalent methods for the detection of parasite larvae in samples of carcasses to be analyzed in a laboratory designated by the competent authority, as reported by Regulation (EC) No 882/2004 [17]. These tests are generally performed by the Istituto Zooprofilattico Sperimentale (IZS) located in different Italian regions, even if also other laboratories may conduct the analysis for official controls as long as they are accredited in accordance to European standards. Such accreditation results from a conformity assessment aiming at verifying whether specific processes, systems, personnel, or organizations comply with specific requirements. Nowadays, the global accreditation system is formed by the International Laboratory Accreditation Cooperation (ILAC), including 120 accreditation bodies across the globe with the purpose to promote the coordination of the accreditation activities [18]. In Italy, the only appointed Italian Accreditation Authority is ACCREDIA since December 2009 in accordance with Regulation (EC) No 765/2008 [19]. It evaluates the technical competence of operators and the credibility of the attestations they release [20]. The aim of this study is the description of the accreditation procedure followed by the internal laboratory attached to an abattoir located in Marche region (Central Italy) for *Trichinella* spp. detection in swine carcasses after the slaughtering process. To our knowledge, this laboratory is one of the only 16 national accredited laboratories placed inside slaughterhouses. The requirements for the maintenance of such accreditation are also reported.

2. Materials and Methods

2.1. Preliminary Information

The accreditation procedure described in the present study was assessed in accordance with EN International Organization for Standardization/International Electrotechnical Commission (ISO/IEC) 17025/2005 on "General requirements for the competence of testing and calibration laboratories" except for sampling made by the competent authority and analytical technique following the official method [16]. The internal laboratory analyzed carcasses of domestic swine coming from holdings not officially recognized as applying controlled housing conditions, as well as carcasses of horses, wild boar, and other farmed and wild animal species susceptible to *Trichinella* spp. infestations. The organization chart is reported in Figure 1. By way of derogation from Article 3, paragraph 3 of Commission Regulation (EU) No 1375/2015 [16], carcasses and meat of domestic swine may be exempt from *Trichinella* spp. examination, as reported also by a national document of the Ministry of Health [21], provided that no autochthonous *Trichinella* spp. infestations have been detected in the last three years. Moreover, another national document of the Ministry of Health [22] specifies that the monitoring of *Trichinella* spp. regarding 10% of carcasses is not necessary anymore on the basis of the epidemiological Italian status but remains mandatory only for carcasses of breeding sows and boars.

The EN ISO/IEC 17025/2005 permits laboratories (public or private, managed by government, industry, or other organizations) to demonstrate that they work expertly and generate valid results that can be accepted by other countries without further analysis. It has been recently revised by EN ISO/IEC 17025/2017, and ILAC in consultation with ISO agreed that a three-year period from the date of publication should be granted to laboratories needing to transition their processes to this new version. During such transition period, both ISO/IEC 17025/2005 and ISO/IEC 17025/2017 could be considered equally valid and applicable [23].

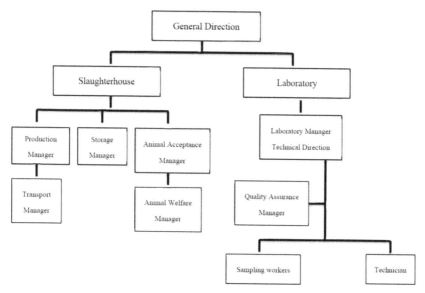

Figure 1. Organization chart reported in the Quality Assurance Manual.

2.2. Preparation of the Quality Assurance Manual

The preparation of a Quality Assurance Manual (QAM) was the first step for the accreditation. It can be considered the reference manual for personnel working in the slaughterhouse, animal owners as customers of the abattoir, and ACCREDIA. It defines the quality policy and organization of the company, the procedures, the responsibilities, as well as the methods of performance. In particular, the quality policy, prepared by the Technical Direction of the Laboratory, involved the evaluation of customer's requests, data obtained, and prospective skills. It was firstly described to the personnel during a meeting and then exposed on a bulletin board. In this context, a Quality Assurance Manager was appointed to check the achievement of such objectives and establish operating procedures and implementation times. The General Direction ensured the availability of adequate resources to obtain the defined skills and verified the whole Quality Assurance System (QAS) whenever a modification of one procedure was required.

2.3. First Accreditation

The QAM was evaluated by ACCREDIA, and subsequently two auditors visited the laboratory and carefully observed the trial carried out by the internal personnel. After their approval, they conferred an accreditation certificate with a validity for four years.

2.4. Maintenance of the Accreditation Certificate

The inspection from ACCREDIA, as described before, was repeated every four years in order to obtain the accreditation renewal, whereas a surveillance assessment was performed each year to verify that the accredited laboratory continued to meet the requirements throughout the validity period of the accreditation certificate. In addition, a proficiency testing according to Marucci et al. [24] was made once a year on meatballs from female mice artificially infected with *T. spiralis* muscle larvae by the European Union Reference Laboratory for Parasites (EURLP, i.e., the Istituto Superiore di Sanità in Rome, Italy). In particular, the proficiency samples could be used to demonstrate performance at an adequate level of sensitivity and, in such occasion, they were spiked with a low number of larvae (3–5), whereas a higher number of larvae could be suitable for training and troubleshooting purposes. The results of the proficiency testing were good when all added larvae were found and acceptable when at least one larva was identified from each spiked sample. The test failed when no larvae was recovered, and in such case, the analyst competence had to be investigated by reviewing data records and/or retesting [25]. The failure of the proficiency testing could cause the suspension of accreditation until a new compliant result was obtained.

The operators working in the accredited laboratory required one spiked sample to the EURLP twice a year only for a training purpose, but these results did not affect the accreditation validity.

2.5. Reference Method of Trichinella spp. Detection

According to Commission Regulation (EU) No. 1375/2015 [16], the magnetic stirrer method for pooled sample digestion was applied. Briefly, 100 g of pooled samples taken from a pillar of the diaphragm of swine carcasses was chopped in a blender containing water, pepsin, and hydrochloric acid. The digestion fluid was stirred until the meat particles disappeared and then poured through a sieve into a sedimentation funnel for 30 min. A portion (40 mL) of the digestion fluid was run off into a measuring cylinder for 10 min, and then 30 mL of supernatant was withdrawn. The remaining 10 mL of sediment was poured into a larval counting basin or Petri dish and examined by trichinoscope or stereo-microscope at a 15 to 20× magnification.

3. Results and Discussion

The results of the present study show the accreditation procedure of the internal laboratory attached to a slaughterhouse in 2012 for the first time and its maintenance up to the current year. The description of the whole process is reported in Table 1, while Table 2 shows QAM, consisting of 19 procedures. Such laboratory was accredited for the analysis of carcasses of swine and other species susceptible to *Trichinella* spp. infection obtained in the slaughterhouse, with an extension also for the examination of carcasses of large wild game and pigs slaughtered at farm level. In order to maintain such accreditation, the technician should: (i) Participate in a training course (at least 8 h once a year), (ii) Pass an annual inter-laboratory test, and (iii) Practice with infected meatball samples every six months. The QAM was revised in four occasions, three of which were due to non-compliances revealed by ACCREDIA, such as the lack of the calibration certificate of a specific instrument, the duplication of some equipment to work faster, or the failure to complete an acceptance report for pigs slaughtered at home. The last revision was made in 2016 after the repealing of Commission Regulation (EC) No 2075/2005 [14] by Commission Regulation (EU) No 1375/2015 [16].

Table 1. Description of the activities for the accreditation.

Process	Responsible	Input	Output	Activities
Customer management	QAM	Sample identification	Sample to be analyzed	Draft acceptance sheet
Supplying	TD	Stock management, calibration plan	Products and services	Selection and control of suppliers and products
Sampling	QAM	Competent authority	Laboratory	Sampling from carcasses and identification
Analysis	TD	Sample delivery	Test report	Application of analytical method
Result communication	TD	Test report	Customer and/or competent authority delivery	Registration of delivery and original copy
Resource management	TD	Maintenance, setting and operator qualification plan	Setting, training, proficiency testing, and quality control	Ordinary and special maintenance, setting of equipment, personal training
Revision	GD	Documents	Review report and improvement actions	Process performance analysis

Legend: QAM = Quality Assurance Manager, TD = Technical Direction, GD = General Direction.

Table 2. List and description of procedures of the Quality Assurance Manual.

Number	Title	Scope and Application
P1	Protection of confidential information	The laboratory guarantees the confidentiality of customer data
P2	Management of documentation of Quality Assurance System	The external or internal documents are identified, registered, and listed in a specific archive
P3	Definition of the contract	The customers accept and sign the proposal of the contract with the Laboratory
P4	Selection of suppliers	The Laboratory selects a specific list for qualified suppliers of reagents, instruments, and equipment and periodically updates it
P5	Check of reagents and equipment	The Laboratory verifies the correct status of equipment and expiry date and the suitability for use of reagents (e.g., pepsine) for the analysis
P6	Resolution of customers' complaints	The customers' complaints are registered in a database and analyzed in order to be solved by corrective actions
P7	Management of non-compliance	If the Laboratory verifies a non-compliance, it prepares a document named "Risk Assessment" with corrective actions to be applied
P8	Management of corrective and preventive actions	The Laboratory evaluates the cause of non-compliance and implements corrective/preventive actions in order to avoid the recurring of such event
P9	Management of audit	The audits are conducted annually aiming at verifying if the Quality Assurance System is compliant with ISO/IEC 17025:2005
P10	Revision by General Direction	The General Direction makes an annual revision of the Quality Assurance System generally after the visit by ACCREDIA
P11	Training of personnel	The Technical Direction guarantees the training of personnel and confirms its effectiveness
P12	Monitoring and maintenance of environmental conditions	The procedures of sanitation are applied daily after the analysis and more accurately each month

Table 2. *Cont.*

Number	Title	Scope and Application
P13	Validation of analysis	The validation involves only the good practice of both equipment and technician because the applied method is already officially recognized
P14	Control of data	The Technical Direction examines and signs the test report
P15	Management of equipment	The Laboratory tests the equipment before the analysis in order to check its suitability
P16	Management and setting of instruments	The Laboratory performs the setting of instruments when required
P17	Sampling	The sampling is carried out by qualified personnel under the supervision of the competent authority
P18	Handling and storage of sample	The Laboratory controls the integrity, temperature, and quantity of samples at their arrival
P19	Quality Assurance	The Laboratory assures the quality of the analysis by using spiked meatballs as a reference certified material and participating in inter-laboratory tests

The accreditation procedure described in this study involved some difficulties for the technician working in the laboratory, such as the structural features required by ACCREDIA and the training courses to be followed in order to maintain the accredited status. Such courses were rarely organized by IZS, even if a minimum number was not mandatory according to ACCREDIA auditors, and therefore the laboratory could consider performing this task.

Another complication was related to the potential failing of the proficiency testing for both the maintenance of accreditation and the training of personnel. In the first case, besides a detailed report of this failure to provide to ACCREDIA auditors, other three spiked meatballs should be purchased from EURLP by the technician, with an expensive charge for the laboratory. For a training purpose, just one spiked sample was considered sufficient. In the present study, the results of the proficiency testing were always acceptable because 2–3 out of four larvae (spiked samples) were detected. This quantitative evaluation was based on the Z-score established by EURLP.

The management of non-compliance and corrective actions (Procedures P7 and P8 reported in Table 2) made by the QAM is shown in Figure 2. Such non-compliance could be observed both into the logistic cycle of the laboratory (i.e., sampling, transport, storage, analysis, and result communication) and into the consideration of prescriptive standards reported in QAM.

The advantages of the accreditation of the internal laboratory are represented by the opportunity to get results quickly (2–3 h after slaughtering), whereas if the analysis was made by the official laboratory (i.e., IZS) the results would be delivered after many hours or even the day after the slaughtering. For this reason, the customer owner of slaughtered animals prefers to pay the analysis made by the internal laboratory in order to obtain fast results. Moreover, because of the extension of accreditation also for the analysis of swine carcasses slaughtered at farm level, as well as large wild game, their meat could be manufactured and processed after a short time from slaughtering. Indeed, when the analyses of pigs slaughtered at farm level were made by IZS, the results were provided with a large delay because the competent authority collected the samples only twice a week and delivered them late.

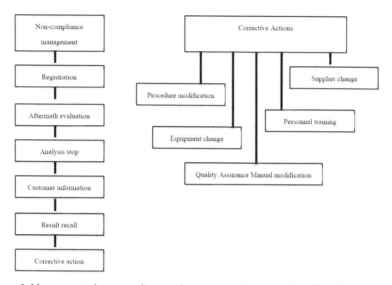

Figure 2. Management of non-compliance and corrective actions (procedures P7 and P8 in Table 2).

Active *Trichinella* spp. control is essential for ensuring food safety and facilitating international trade. Additionally, for pork carcass safety assurance, considering *Trichinella* spp. risk assessment, the effective control of this hazard depends on preventive measures and controls continuously applied on farm and at the slaughterhouse in an integrated way [26]. The European Legislation requires that carcasses of animal susceptible to such parasite must be examined before they can be intended for human consumption. The possibility of doing such analyses inside a slaughterhouse is useful for customers as well as for the competent authority in guaranteeing the safety of the slaughtered carcasses. Periodic reviews of the entire accreditation procedure must be conducted and planned at least every five years in order to certify the reliability of the results, even if the experience has shown that it represents a complex and laborious process.

Author Contributions: For research articles with several authors, a short paragraph specifying their individual contributions must be provided. The following statements should be used conceptualization, M.S. and P.V.; methodology, A.M.A.O.; formal analysis, M.P.S.; writing—original draft preparation, M.S. and P.V.; writing—review and editing, M.S., P.V., and A.P.; supervision, A.P.; funding acquisition, A.M.A.O. and M.P.S.

Funding: This research received no external funding.

Conflicts of Interest: The authors declare no conflict of interest.

References

1. Dimzas, D.; Diakou, A.; Koutras, C.; Gómez-Morales, M.A.; Psalla, D.; Keryttopoulos, P.; Deligianni, D.; Kontotasios, K.; Pozio, E. Human trichinellosis caused by *Trichinella britovi* in Greece, and literature review. *J. Helminthol.* **2019**, *13*, 1–4. [CrossRef] [PubMed]
2. Rostami, A.; Gamble, H.R.; Dupouy-Camet, J.; Khazan, H.; Bruschi, F. Meat sources of infection for outbreaks of human trichinellosis. *Food Microbiol.* **2017**, *64*, 65–71. [CrossRef] [PubMed]
3. Gómez-Morales, M.A.; Ludovisi, A.; Amati, M.; Cherchi, S.; Tonanzi, D.; Pozio, E. Differentiation of *Trichinella* species (*Trichinella spiralis*/*Trichinella britovi* versus *Trichinella pseudospiralis*) using western blot. *Parasite Vectors* **2018**, *11*, 1–10. [CrossRef] [PubMed]
4. Turiac, I.A.; Cappelli, M.G.; Olivieri, R.; Angelillis, R.; Martinelli, D.; Prato, R.; Fortunato, F. Trichinellosis outbreak due to wild boar meat consumption in southern Italy. *Parasite Vectors* **2017**, *10*, 1–2. [CrossRef] [PubMed]

5. Pozio, E. Adaptation of *Trichinella* spp. for survival in cold climates. *Food Waterborne Parasitol.* **2016**, *4*, 4–12. [CrossRef]

6. Zarlenga, D.; Wang, Z.; Mitreva, M. *Trichinella spiralis*: Adaptation and parasitism. *Vet. Parasitol.* **2016**, *231*, 8–21. [CrossRef] [PubMed]

7. Saracino, M.P.; Calcagno, M.A.; Beauche, E.B.; Garnier, A.; Vila, C.C.; Granchetti, H.; Taus, M.R.; Venturiello, S.M. *Trichinella spiralis* infection and transplacental passage in human pregnancy. *Vet. Parasitol.* **2016**, *231*, 2–7. [CrossRef] [PubMed]

8. Xu, J.; Yang, F.; Yang, D.Q.; Jiang, P.; Liu, R.D.; Zhang, X.; Cui, J.; Wang, Z.Q. Molecular characterization of *Trichinella spiralis* galectin and its participation in larval invasion of host's intestinal epithelial cells. *Vet. Res.* **2018**, *49*, 1–15. [CrossRef] [PubMed]

9. Park, M.K.; Kang, Y.J.; Jo, J.O.; Baek, K.W.; Yu, H.S.; Choi, Y.H.; Cha, H.J.; Ock, M.S. Effect of muscle strength by *Trichinella spiralis* infection during chronic phase. *Int. J. Med. Sci.* **2018**, *15*, 802–807. [CrossRef] [PubMed]

10. Dupouy-Camet, J.; Kociecka, W.; Bruschi, F.; Bolas-Fernandez, F.; Pozio, E. Opinion on the diagnosis and treatment of human trichinellosis. *Expert Opin. Pharmacother.* **2002**, *3*, 1117–1130. [PubMed]

11. Gottstein, B.; Pozio, E.; Nöckler, K. Epidemiology, diagnosis, treatment and control of trichinellosis. *Clin. Microbiol. Rev.* **2009**, *22*, 127–145. [CrossRef] [PubMed]

12. European Food Safety Authority and European Centre for Disease Prevention and Control (EFSA and ECDC). The European Union summary report on trends and sources of zoonoses, zoonotic agents and food-borne outbreaks in 2017. *EFSA J.* **2018**, *16*, e05500.

13. Badagliacca, P.; Di Sabatino, D.; Salucci, S.; Romeo, G.; Cipriani, M.; Sulli, N.; Dall'Acqua, F.; Ruggieri, M.; Calistri, P.; Morelli, D. The role of the wolf in endemic sylvatic *Trichinella britovi* infection in the Abruzzi region of Central Italy. *Vet. Parasitol.* **2016**, *231*, 124–127. [CrossRef] [PubMed]

14. Commission Regulation (EC) No 2075/2005 of 5 December 2005 Laying Down Specific Rules on Official Controls for Trichinella in Meat. Available online: https://eur-lex.europa.eu/LexUriServ/LexUriServ.do?uri= OJ:L:2005:338:0060:0082:EN:PDF (accessed on 5 June 2019).

15. Franssen, F.; Swart, A.; van der Giessen, J.; Havelaar, A.; Takumi, K. Parasite to patient: A quantitative risk model for *Trichinella* spp. pork and wild boar meat. *Int. J. Food Microbiol.* **2017**, *241*, 262–275. [CrossRef] [PubMed]

16. Commission Implementing Regulation (EU) No 1375/2015 of 10 August 2015 Laying Down Specific Rules on Official Controls for Trichinella in Meat. Available online: https://eur-lex.europa.eu/legal-content/EN/TXT/ PDF/?uri=CELEX:32015R1375&from=EN (accessed on 5 June 2019).

17. Regulation (EC) No 882/2004 of the European Parliament and of the Council of 29 April 2004 on Official Controls Performed to Ensure the Verification of Compliance with Feed and Food Law, Animal Health and Animal Welfare Rules. Available online: http://www.cefas.co.uk/media/1550/extract_reg_no_882_2004.pdf (accessed on 5 June 2019).

18. Zhai, P.; Wang, R.; Zhou, Y.; Hu, D.; Ii, J.; Zhou, Y. Enhancing the capabilities of biosafety laboratories through the established accreditation system: Development of the biosafety laboratory accreditation system in China. *J. Biosaf. Biosecur.* **2019**. [CrossRef]

19. Regulation (EC) No 765/2008 of the European Parliament and of the Council of 9 July 2008 Setting out the Requirements for Accreditation and Market Surveillance Relating to the Marketing of Products and Repealing Regulation (ECC) No 339/1993. Available online: https://eur-lex.europa.eu/LexUriServ/LexUriServ. do?uri=OJ:L:2008:218:0030:0047:EN:PDF (accessed on 5 June 2019).

20. Ricci, U.; De Sanzo, C.; Carboni, I.; Iozzi, S.; Nutini, A.L.; Torricelli, F. Accreditation of a forensic genetics laboratory in Italy. *Forensic Sci. Int.: Gen. Suppl. Ser.* **2013**, *4*, e294–e295. [CrossRef]

21. Ministry of Health. Esame Trichinoscopico Delle Carni di Suini Domestici—Deroga ai Sensi del Regolamento (UE) No. 1375/2015. Available online: https://www.anmvioggi.it/images/CIRCOLARE_DEL_MINISTERO_ DELLA_SALUTE_copy_copy.pdf (accessed on 5 June 2019).

22. Ministry of Health. Monitoraggio per il Controllo Delle Trichine ai Sensi Dell'articolo 11 del Regolamento (UE) No 1375/2015. Available online: http://www.ulss4.veneto.it/web/ulss4/Prevenzione/dfsa/normative/norme_ nazionali/note_circolari_minsal/0_pagina_iniziale/all/2016/monit_contr_trichine_Reg_UE_2015_1375.pdf (accessed on 5 June 2019).

23. Joint ILAC/ISO Communiqué on the Recognition of ISO/IEC 17025 during a Three-Year Transition, 01 November 2017. Available online: http://www.esyd.gr/pweb/s/20/files/anakoinoseis/Joint_ILAC-ISO_Com_-_ISO_IEC_17025_transition.pdf (accessed on 5 June 2019).

24. Marucci, G.; Tonanzi, D.; Cherchi, S.; Galati, F.; Bella, A.; Interisano, M.; Ludovisi, A.; Amati, A.; Pozio, E. Proficiency testing to detect *Trichinella* larvae in meat in the European Union. *Vet. Parasitol.* **2016**, *231*, 145–149. [CrossRef] [PubMed]

25. International Commission on Trichinellosis. Recommendations for Quality Assurance in Digestion Testing Programs for *Trichinella*, ICT Quality Assurance Committee (Appendix 1) Part 3, 2012; pp. 1–9. Available online: http://www.trichinellosis.org/uploads/PART_3__final__-_PT_7Feb2012.pdf (accessed on 5 June 2019).

26. EFSA Panels on Biological Hazards (BIOHAZ), on Contaminants in the Food Chain (CONTAM), and on Animal Health and Welfare (AHAW). Scientific Opinion on the public health hazards to be covered by inspection of meat (swine). *EFSA J.* **2011**, *9*, 2351.

Article

Validation Including Uncertainty Estimation of a GC–MS/MS Method for Determination of Selected Halogenated Priority Substances in Fish Using Rapid and Efficient Lipid Removing Sample Preparation

Slávka Nagyová and Peter Tölgyessy *

Water Research Institute, Slovak National Water Reference Laboratory, Nábrežie arm. gen. L. Svobodu 5, 812 49 Bratislava, Slovak Republic
* Correspondence: peter.tolgyessy@vuvh.sk; Tel.: +421-2-59343-466

Received: 20 February 2019; Accepted: 13 March 2019; Published: 18 March 2019

Abstract: A rapid method is proposed for the determination of selected H_2SO_4 stable organic compounds—eight organochlorines (OCs; hexachloro-1,3-butadiene, pentachlorobenzene, hexachlorobenzene, hexachlorocyclohexane—HCH—isomers, heptachlor) and six polybrominated diphenyl ethers (PBDEs; BDE-28, 47, 99, 100, 153, 154)—in fish samples. In the method, a modified QuEChERS (quick, easy, cheap, effective, rugged and safe) sample preparation using pH-tuned dispersive liquid–liquid microextraction (DLLME) and H_2SO_4 digestion fish extract clean-up is followed by gas chromatography–triple quadrupole tandem mass spectrometry (GC–QqQ-MS/MS) analysis. The method was validated in terms of linearity, limits of the method, recovery, accuracy, analysis of standard reference material (NIST SRM 1946), and estimation of combined uncertainty of the measurement (top-down approach). For validation, chub composite samples were used, and subsequently, the method was successfully applied to analysis of real samples of eight fish species. Finally, the method passed the analytical Eco-Scale evaluation as "an acceptable green analysis method", and showed its advantages (simplicity, rapidity, low cost, high extract clean-up efficiency, good sensitivity) when compared to other reported QuEChERS based methods.

Keywords: QuEChERS; dispersive liquid–liquid microextraction; sulfuric acid treatment; gas chromatography; tandem mass spectrometry; priority substances; fish samples

---

1. Introduction

Anthropogenic halogenated organic compounds synthesized as pesticides, solvents or fire retardants have been found to pose a serious threat to aquatic environments, wildlife and humans due to their toxic, persistent and bioaccumulative properties [1]. Because of these harmful effects and impacts, the production and use of large number of organochlorine (OC) pesticides and certain brominated flame retardants was banned or severely restricted in the European Union (EU), and other parts of the world [2], but their presence and release in the environment can be expected over the next decades.

One of the ways human health can be endangered by these substances is through consuming fish living in contaminated waters and accumulating the toxic chemicals in their tissues. Therefore, it is necessary to monitor and analyze fish contamination to protect humans from the consumption of contaminated food. The regulatory limit applicable to residues of pesticides in fish and fishery products is the default maximum residue level (MRL) of 10 µg/kg set by the EU in Regulation 396/2005, which concerns public health and is relevant to the functioning of the internal market [3]. For the determination of PBDEs in fish and other seafood, the EU Commission recommends (recommendation

2014/118/EU) use of analytical methods with a limit of quantification of 0.01 µg/kg wet weight or lower [4].

This paper is focused on the determination of eight OC compounds (hexachloro-1,3-butadiene, pentachlorobenzene, hexachlorobenzene, hexachlorocyclohexane—HCH—isomers, heptachlor) and six polybrominated diphenyl ethers (PBDEs; BDE-28, 47, 99, 100, 153, 154) in fish that were selected from the EU list of priority substances in the field of water policy [5] and from the U.S. EPA (Environmental Protection Agency) priority pollutants list [6]. A great variety of extraction techniques have been applied in the analysis of organic halogenated compounds in fish samples. Among others, solid–liquid extraction (SLE), Soxhlet extraction, accelerated solvent extraction (ASE), supercritical fluid extraction (SFE), microwave-assisted extraction (MAE), matrix solid-phase dispersion (MSPD), and so-called QuEChERS method have been reported in literature [7–10]. The presented methods include both traditional extraction methods (SLE, Soxhlet) which are quite laborious, time-consuming (extraction duration up to 24 h), and require large amounts of organic solvents (up to few hundreds of mL), and novel methods with shortened extraction times (to 10–60 min), reduced solvent consumption, often with reduced cost, and that are amenable to automation. However, the disadvantage of ASE, SFE, and MAE methods lies in the cost of equipment setup.

In the last few decades, the QuEChERS method with the advantages summarized in its acronym (quick, easy, cheap, effective, rugged, and safe) has become a very attractive sample preparation method in food analysis [11]. Overall, this procedure consists of two main parts: extraction with a solvent (mostly acetonitrile, MeCN) and partitioning salts ($MgSO_4$, NaCl), and extract clean-up using dispersive solid-phase extraction (dSPE) technique. However, the dSPE clean-up is not fully sufficient for the analysis of high fat matrices and, therefore, the clean-up part of the original QuEChERS method has gone through various modifications to enhance the co-extractives (mainly lipid) removal efficiency (use of freezing, dual dSPE, gel permeation chromatography, silica minicolumn, EMR-lipid sorbent) [9,12–17].

Currently, a novel method for clean-up of fatty MeCN extracts (after QuEChERS extraction), which is suitable for determination of H_2SO_4 stable organic compounds in complex biological samples, was developed in our laboratory [18]. The sample extract clean-up combines the pH-tuned dispersive liquid–liquid microextraction (DLLME) with conc. H_2SO_4 digestion. This clean-up offers many advantages, including high lipid removal efficiency, rapidity, analyte enrichment without evaporating solvent, low cost (cheap chemicals, no need for expensive sorbents), low chemicals and glassware usage, no need of special laboratory equipment, and less bench space. The lipid removal involves complete removal of fatty acids, which are partitioned from the organic phase into the alkaline aqueous phase in the DLLME clean-up step [18]. The disadvantage is the use of toxic and hazardous chemicals ($CHCl_3$, hexane, MeCN, H_2SO_4), however, they are applied in small quantities.

The aim of this study was the validation including uncertainty estimation of a rapid and non-laborious method for determination of selected H_2SO_4 stable halogenated priority substances in fish employing modified QuEChERS sample preparation followed by gas chromatographic and triple quadrupole tandem mass spectrometric (GC–QqQ-MS/MS) analysis.

2. Materials and Methods

2.1. Standards and Reagents

Neat standards of pentachlorobenzene, hexachlorobenzene, *alpha*-HCH, *beta*-HCH, *delta*-HCH and heptachlor (purity: 98.1–99.5%) were obtained from Dr. Ehrenstorfer (Augsburg, Germany). Neat standards of hexachloro-1,3-butadiene (96%) and lindane (97%) were from Sigma–Aldrich (Steinheim, Germany). Standard of 2,4,5,6-tetrachloro-*m*-xylene (99.0%) in cyclohexane at 10 µg/mL was prepared by Dr. Ehrenstorfer. Individual PBDE standards BDE-28, BDE-47, BDE-77, BDE-99, BDE-100, BDE-153 and BDE-154, each at 50 µg/mL in nonane (≥98%), were produced by Cambridge Isotope Laboratories (CIL, Andover, MA, USA).

Anhydrous magnesium sulfate, sulfuric acid, acetone, chloroform and toluene, all Emsure grade, cyclohexane and ethyl acetate (SupraSolv), and n-hexane (UniSolv), were purchased from Merck (Darmstadt, Germany). Sodium chloride, anhydrous sodium acetate (both ReagentPlus) and MeCN (Chromasolv) were obtained from Sigma–Aldrich.

Sodium acetate solution at 0.5 M was prepared by dissolution of CH_3COONa in Milli-Q water produced by a Direct-Q 3 water purification system (Millipore, Molsheim, France). Stock solutions of each OC compound obtained as a neat material were prepared in cyclohexane at a concentration of 5 mg/mL, with the exception of a solution of *beta*-HCH, which was prepared in a mixture of cyclohexane and acetone (4:1, *v/v*) at a concentration of 1 mg/mL. Standard working mixtures of eight OC compounds were prepared by dilution of their stock solutions with cyclohexane to obtain concentrations of 1 and 10 µg/mL. An internal standard (IS) solution of 2,4,5,6-tetrachloro-*m*-xylene at 1 µg/mL was prepared by dilution of the stock standard solution with cyclohexane. Standard working mixtures of six PBDEs (BDE-28, 47, 99, 100, 153 and 154) at concentrations of 5 and 0.5 µg/mL were obtained by dilution of the individual standard solutions with toluene. An IS solution of BDE-77 at 5 µg/mL was also prepared from the stock standard solution by dilution with toluene.

2.2. Fish Samples

The proposed method was validated and verified using samples of nine different fish species: European chub (*Squalius cephalus*), crucian carp (*Carassius carassius*), European perch (*Perca fluviatilis*), northern pike (*Esox lucius*), zander (*Sander lucioperca*), brown trout (*Salmo trutta*), Atlantic salmon (*Salmo salar*), Alaska pollock (*Theragra chalcogramma*), and lake trout (*Salvelinus namaycush*). The first six species were collected by electrofishing during a fish survey performed in Slovak water bodies in 2015 within the project: Monitoring and assessment of water body status (see Funding). The samples of salmon and pollock were purchased as frozen skinless fillets from a local supermarket. The samples were prepared as composite homogenates from several pieces (from 2 to 7) of the whole fish (chub, perch, pike, and trout) or homogenates from single fish (chub and remaining species). The samples were homogenized using a knife mill Grindomix GM 200 (Retsch, Haan, Germany) to give a wet weight of about 600 g and were stored in a freezer at −20 °C until extraction and analysis. The main part of the study was done using chub composite samples.

Accuracy of the method was demonstrated by the analysis of the standard reference material SRM 1946 (Lake Superior Fish Tissue) which was prepared from lake trout by the National Institute of Standards and Technology (NIST, Gaithersburg, MD, USA). This SRM was a frozen fish tissue homogenate with 10.2% of extractable fat and 71.4% of water.

2.3. Lipid and Moisture Determination

The lipid and moisture content of fish homogenate samples was determined by gravimetric methods according to our work [19]. For total lipid determination, 5 g of fish homogenate was extracted with 5 mL of acetone/ethyl acetate solvent mixture (6:4, *v/v*) by shaking with a vortex mixer (Stuart SA8, Bibby Scientific, Stone, UK) for 3 min and, after addition of 2 g of $MgSO_4$ and 0.5 g NaCl and shaking for 3 min, the organic phase was separated by centrifugation (centrifuge Rotina 380, Hettich, Tuttlingen, Germany). An aliquot of the organic phase was dried to constant weight at 103 °C, and the percent lipid content was calculated from the mass of the final residue. The moisture content was determined from the mass difference of 2–3 g portions of fish homogenate before and after a 24 h drying at 60 °C. For all fish sample homogenates, the lipid content was determined in triplicates (results in the range 0.63–16%) and the moisture content in duplicates (58–81%).

2.4. Sample Preparation

An aliquot of 5 g of fish homogenate was weighed into a 50-mL polypropylene centrifuge tube (Corning, CentriStar, Sigma–Aldrich, Steinheim, Germany) and spiked with IS solutions and mixture of analytes (in case of standard addition). After 15 min, the spiked homogenate was mixed with 5 mL of MeCN and shaken by a vortex mixer at 800 rpm for 1 min. Then, a salt mixture of 2 g of anhydrous $MgSO_4$ and 0.5 g of NaCl was added, and again, the tube was shaken vigorously for 1 min. Next, the sample was centrifuged at 5000 rpm for 5 min.

In the DLLME step, a 1-mL aliquot of supernatant was transferred to a 15-mL centrifuge tube with 4 mL of 0.5 M CH_3COONa solution. Then, 50 μL of $CHCl_3$ was injected rapidly into the mixture; the tube was vortexed for 1 min and centrifuged at 5000 rpm for 5 min.

Finally, for the H_2SO_4 clean-up step, the whole sedimented phase was placed in a 1.7-mL clickseal microcetrifuge tube (GoldenGate Bioscience, Claremont, CA, USA) and 1 mL of concentrated H_2SO_4 was added slowly. The tube was sealed, shortly shaken by hand and then 80 μL of hexane was added to the top of the solution. After short shaking, the tube was centrifuged in a microcentrifuge (Mikro 220R, Hettich) at 10,000 rpm for 5 min. The upper phase was transferred into a GC vial equipped with a 100-μL glass insert and was then ready for GC–MS/MS analysis.

2.5. Instrumental Analysis

Analyses were performed using an Agilent 7890B GC combined with a 7000D QqQ-MS/MS system (Agilent Technologies, Wilmington, DE, USA). The GC was equipped with a multimode inlet and for the injection of sample extracts a multipurpose sampler (MPS) from Gerstel (Mülheim a/d Ruhr, Germany) was used. Two identical Agilent HP-5MS UI capillary columns (15 m × 0.25 mm I.D., 0.25 μm film thickness) connected in series (via Agilent Purged Ultimate Union) were used for separation of the analytes, while a deactivated fused-silica tube (1 m × 0.32 mm I.D.) was used as a precolumn. Helium was used as the carrier gas at constant flow rates of 1.1 and 1.3 mL/min for the first and the second column, respectively.

Sample injection (1 μL) was carried in splitless mode (1 min) at 275 °C. The oven temperature was programmed from 60 °C (1 min hold) to 170 °C at a rate of 40 °C/min, and then to 300 °C (1.75 min hold) at a rate of 10 °C/min. After each run, a 3 min column clean-up was performed employing a mid-column backflush. The backflush was conducted at 305 °C, by applying helium to purged ultimate union at 320 kPa. This program resulted in a total run time of 21.5 min.

The mass selective detector (MSD) was operated using electron ionization at 70 eV in the multiple reaction monitoring (MRM) mode. The retention times (t_R), quantifier and qualifier transitions for the selected analytes are listed in Table 1. Dwell times were in all cases set at 10 ms. The MSD transfer line was at 280 °C, ion source at 300 °C, and quadrupoles at 150 °C. The QqQ collision gas was nitrogen at 1.5 mL/min, and quench gas was helium at 2.25 mL/min. Agilent MassHunter software was used for instrument control and data analysis.

Table 1. Analytes, retention times and MRM conditions.

Analyte	t_R (min)	MRM Transitions (m/z)			
		Quantifier	CE (V)	Qualifier	CE (V)
Hexachloro-1,3-butadiene	4.42	225→190	15	260→225	15
Pentachlorobenzene	6.12	248→213	25	250→180	20
Tetrachloro-*m*-xylene (IS-1)	6.79	244→209	15	171→136	15
alpha-HCH	7.36	219→183	5	217→181	15
Hexachlorobenzene	7.50	284→214	35	284→249	20
beta-HCH	7.75	219→183	5	217→181	15
Lindane	7.86	219→183	5	217→181	15
delta-HCH	8.22	219→183	5	217→181	15
Heptachlor	9.02	272→237	25	272→117	35
BDE-28	11.94	246→139	30	406→246	20
BDE-47	14.02	326→217	30	486→326	20
BDE-77 (IS-2)	14.74	326→217	30	486→326	20
BDE-100	15.55	564→404	20	404→297	30
BDE-99	15.97	564→404	20	404→297	30
BDE-154	17.19	644→484	20	484→324	40
BDE-153	17.84	644→484	20	484→324	40

Abbreviations: t_R—retention time; MRM—multiple reaction monitoring; CE—collision energy.

The quantification process was performed using a single point standard addition method applying Equation (1):

$$c_i = c_{ad} \times \frac{A_i / A_{IS}}{(A_{i+ad}/A_{ISsa}) - (A_i/A_{IS})} \qquad (1)$$

where c_i is the determined analyte concentration, c_{ad} is the added concentration to the sample, A_i and A_{IS} are the peak areas of the analyte and IS from the unknown sample analysis, A_{i+ad} and A_{ISsa} are the peak areas of the analyte and IS from the analysis with standard addition. For this purpose, the concentration of each added analyte and of IS tetrachloro-*m*-xylene was 10 µg/kg and of IS BDE-77 was 20 µg/kg. This was appropriate for the studied range and in agreement with the study of Frenich et al. [20]. In the whole work, the concentrations of the analytes are presented on a wet weight basis.

2.6. Matrix Effect Evaluation

The evaluation of the matrix effect (*ME*) on the GC–MS/MS analysis was based on comparing the analyte response measured in matrix-matched extracts spiked after QuEChERS extraction and processed by DLLME and H_2SO_4 clean-up procedure and the response measured in a corresponding neat solvent solution. The *ME* was calculated from replicate analyses as the average percent suppression or enhancement in the peak area using the following Equation (2):

$$ME(\%) = \frac{Peak\ area\ in\ matrix\ matched\ standard - Peak\ area\ in\ solvent\ standard}{Peak\ area\ in\ matrix\ matched\ standard} \times 100 \qquad (2)$$

A positive value of *ME* corresponds to a matrix-induced enhancement of analyte response, whereas a negative value corresponds to a suppression effect.

2.7. Measurement Uncertainty Calculation

The combined measurement uncertainty was estimated according to the top-down approach using quality control (QC) charts, validation data and the uncertainty of purity of analytical standards [21,22]. The random error contribution to the measurement uncertainty was characterized by the within-lab (intermediate) reproducibility ($u_{r,repro}$), which was calculated as relative standard deviation (RSD%)

from at least 20 independent consecutive measurement values taken from the QC charts (QC samples spiked at 5 µg/kg).

Systematic components of uncertainty were characterized as the relative bias (B_r) and the uncertainty of the systematic error ($u_{r,cm}$) and were determined by measuring QC samples at conditions of repeatability. The B_r was quantified using the Equation (3):

$$B_r(\%) = \frac{c_m - c_{ref}}{c_{ref}} \times 100 \tag{3}$$

where c_m and c_{ref} are the mean measured concentration and reference concentration of the studied analyte, respectively. For determination of $u_{r,cm}$, the QC samples were analyzed with minimum number of replicates of 10. For calculation, the following Equations (4) and (5) were used:

$$u_{cm} = \frac{SD}{\sqrt{n}} \tag{4}$$

$$u_{r,cm}(\%) = \frac{u_{cm}}{c_{ref}} \times 100 \tag{5}$$

where u_{cm} is the standard uncertainty of c_m, SD is the standard deviation, and n is the number of replicates.

The uncertainty of purity of analytical standards ($u_{r,ref}$) was determined by dividing the expanded combined uncertainty ($U_{r,ref}$) (given in manufacturer's certificate) by the coverage factor $k = 2$, or was estimated on the basis of Equation (6) derived from rectangular distribution (in case of absence of the certificate):

$$u_{r,ref}(\%) = \frac{0.5 \times (100 - y)}{\sqrt{3}} \tag{6}$$

where y (%) represents the purity of standard given in the manufacturer's specification.

All the characterized uncertainty components were combined by the error propagation rule to obtain the relative combined measurement uncertainty ($u_{r,tot}$) using Equation (7):

$$u_{r,tot}(\%) = \sqrt{u_{r,repro}^2 + B_r^2 + u_{r,cm}^2 + u_{r,ref}^2} \tag{7}$$

Finally, the expanded combined uncertainty ($U_{r,tot}$) was calculated by multiplying the $u_{r,tot}$ with the coverage factor of 2 (95% confidence level).

3. Results and Discussion

3.1. Instrumental Analysis

The instrumental analysis conditions summarized in Section 2.5 were selected based on our previous studies [9,18]. In contrast to study [18], in which sample loading for GC analysis was carried out by thermal desorption of fish extract from the insert placed in the thermal desorption tube, a liquid injection of sample extract was employed.

3.2. ME

The QuEChERS extracts from the chub homogenate sample with lipid content of 5.2% and water content of 74.6% spiked with test analytes at concentration level of 5 ng/mL and treated by DLLME and H_2SO_4 were used for ME evaluation according to 2.6. The MEs were calculated from five replicate analyses. The obtained ME values in the range from −5.1% to 10.5% (see Table 2) show very low enhancement or suppression of chromatographic response of the studied analytes. For comparison, MEs presented in the studies employing modified QuEChERS methods with dSPE clean-up were for selected OC pesticides incomparably higher. The ME values, in the studies [14] and [23], were 32 and 175.7% for *beta*-HCH, and 63 and 219.4% for *delta*-HCH, respectively. The *ME*s for *alpha*-HCH, hexachlorobenzene, lindane and heptachlor were in the study [23] calculated as 40.3, 27.4, 34.7 and 28.8%, respectively. For seven PBDEs, Sapozhnikova and Lehotay [24] observed matrix-induced suppression of chromatographic response with *ME* values in the range from −16% to −26% when using unbuffered QuEChERS method with dSPE clean-up for fish sample preparation. The low *ME*s obtained by the proposed method in this study indicate the high efficiency of fish extract clean-up.

Table 2. Matrix effect (*ME*) evaluation for the studied analytes in the spiked QuEChERS extracts after DLLME and H_2SO_4 clean-up ($n = 5$).

Analyte	*ME* (%)	RSD (%)
Hexachloro-1,3-butadiene	−5.1	11
Pentachlorobenzene	−1.9	11
Tetrachloro-*m*-xylene (IS-1)	−1.3	12
alpha-HCH	−1.2	11
Hexachlorobenzene	3.8	13
beta-HCH	9.3	10
Lindane	2.3	10
delta-HCH	3.6	11
Heptachlor	1.0	12
BDE-28	1.6	14
BDE-47	5.7	16
BDE-77 (IS-2)	1.3	15
BDE-100	3.9	16
BDE-99	10	14
BDE-154	7.0	14
BDE-153	11	14

3.3. Method Validation

Within-laboratory validation of the proposed method was carried out using two chub composite samples with lipid and water content of 1.9% and 5.2%, and 80% and 75%, respectively, with absence or low levels of the analytes of interest. The validation was performed in terms of linearity, limits of detection (LOD), limits of quantification (LOQ), recovery, accuracy involving evaluation of precision and trueness and analysis of SRM and, finally, the combined uncertainty of the measurement was estimated.

3.3.1. Linearity

Response linearity was assessed by studying calibration curves from the analyses of matrix-matched standards of the test analytes. The standards were prepared by spiking the extract of chub composite sample (lipid content of 1.9%) with standard working mixtures to obtain seven concentration levels (0.1, 0.5, 1, 5, 15, 30 and 60 μg/kg). The response linearity was evaluated on the basis of coefficients of determination (R^2) and RSDs of the relative response factors (RRF). The RRFs of the analytes were calculated relative to the internal standard at each concentration level applying a blank correction. In Table 3 it can be seen that the obtained calibration functions were linear for all the analytes, with R^2 values above 0.999 and RSDs of the RRFs in the range of 5.6–13%.

Table 3. Linearity, limits and accuracy of the proposed method for determination of test analytes in spiked fish matrix.

Analyte	Linear Range (µg/kg)	R^2	RRF	RRF_RSD (%)	LOD (µg/kg)	LOQ (µg/kg)	Precision		Trueness	
							Pre$_{intra}$	Pre$_{inter}$	R	B$_r$
							RSD (%)	RSD (%)	(%)	(%)
Hexachlorobutatadiene	0.1–60	0.99997	1.7	6.7	0.028	0.092	3.0	5.7	95	−4.7
Pentachlorobenzene	0.1–60	0.99975	1.0	5.8	0.036	0.12	2.3	4.9	95	−5.2
alpha-HCH	0.1–60	0.99989	3.3	7.1	0.029	0.096	0.5	6.7	89	−11
Hexachlorobenzene	0.1–60	0.99986	1.4	5.9	0.052	0.17	0.8	3.2	107	7.4
beta-HCH	0.1–60	0.99973	2.1	11	0.036	0.12	3.4	7.2	88	−12
Lindane	0.1–60	0.99987	2.4	12	0.040	0.13	1.6	7.0	87	−13
delta-HCH	0.1–60	0.99982	2.0	12	0.037	0.12	4.7	8.3	91	−9.4
Heptachlor	0.1–60	0.99952	0.46	6.9	0.039	0.13	9.2	16	94	−6.2
BDE-28	0.1–60	0.99984	5.3	5.6	0.037	0.12	8.9	9.0	99	−1.3
BDE-47	0.1–60	0.99994	3.2	13	0.028	0.092	4.8	6.6	102	1.9
BDE-100	0.1–60	0.99984	1.6	10	0.028	0.092	2.9	9.0	100	0.22
BDE-99	0.1–60	0.99986	1.2	5.9	0.021	0.072	6.6	9.2	99	−0.89
BDE-154	0.1–60	0.99939	0.49	12	0.049	0.16	9.0	13	101	1.2
BDE-153	0.1–60	0.99912	0.26	8.1	0.042	0.14	11	13	105	5.1

Abbreviations: R^2—coefficient of determination; RRF—relative response factor; LOD—limit of detection; LOQ—limit of quantification; PRE$_{intra}$—intra-day precision; PRE$_{inter}$—inter-day precision; R—recovery; B$_r$—relative bias.

3.3.2. Limits of the Method

Ten replicate analysis of the blank chub composite sample spiked at 0.1 µg/kg were used for determination of the limits of the method. The LODs and LOQs were calculated as three and ten times the standard deviations (SD) of the results, respectively. As can be seen in Table 3, the LOQs for the test analytes were in the range from 0.07 to 0.17 µg/kg, much lower than the default MRL of 10 µg/kg applicable to pesticides in fish, and approaching the EU Commission recommended value of 0.01 µg/kg for LOQ of analytical methods for the determination of PBDEs in fish and other seafood [4]. In the published studies [14,23] employing similar GC–MS/MS instrumentation and applying the QuEChERS methodology with dSPE clean-up for determination of the test OC compounds in fish, the LOQs were in the ranges 2–13 µg/kg, and 1–5 µg/kg, respectively. In these studies, unlike with the employed SD calculation approach, a less/not appropriate (for QqQ-MS/MS detection, [22]) method based on signal to noise (S/N) estimation was used for LOQ evaluation, and therefore the obtained LOQs are hardly comparable. The lowest calibration levels (LCLs) of the test PBDEs obtained in the study [24] using original QuEChERS sample preparation and GC–MS/MS method were in the range 0.5–5 µg/kg. For illustration, Figure 1 shows a total MRM chromatogram from the analysis of the blank chub composite sample spiked with the test analytes at LOQ level of 0.1 µg/kg. It can be seen the baseline separation of all the analytes with no interferences from the matrix.

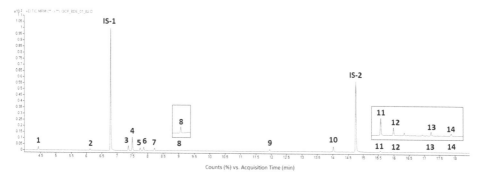

Figure 1. Total ion MRM chromatogram from the GC–QqQ-MS/MS analysis of the blank chub composite sample spiked with the test analytes at LOQ level of 0.1 µg/kg, and internal standards at 10 (IS-1) and 20 µg/kg (IS-2), respectively. Peaks: 1—hexachloro-1,3-butadiene, 2—pentachlorobenzene, IS-1—tetrachloro-*m*-xylene, 3—*alpha*-HCH, 4—hexachlorobenzene, 5—*beta*-HCH, 6—lindane, 7—*delta*-HCH, 8—heptachlor, 9—BDE-28, 10—BDE-47, IS-2—BDE-77, 11—BDE-100, 12—BDE-99, 13—BDE-154, 14—BDE-153. Hexachlorobenzene (at 0.64 µg/kg), BDE-47 (0.35 µg/kg), BDE-100 (0.18 µg/kg), BDE-154 (0.23 µg/kg), and BDE-153 (0.15 µg/kg) were present in the sample before spiking.

3.3.3. Recovery

Recovery experiments were conducted with the chub composite sample (lipid content of 1.9%) spiked with the test analytes at levels of 1, 5, 15, 30 and 60 µg/kg, respectively, covering the linearity range of the method. A single-point standard addition method (see Section 2.5) was used for quantification, which enabled us to solve the problem of absence of suitable fish matrix free of analytes of interest that is necessary for matrix-matched calibration at low concentration levels, and also helped to overcome the negative effects of matrix components. From Table 4 it can be seen that the obtained recoveries are in the range of 57%–124% with RSDs in the range of 2%–18%. These results are acceptable according to the requirements of the EU guidance document SANTE/11813/2017 for pesticide residues analysis in food, because the recoveries outside the range of 70%–120% are consistent (RSDs ≤ 20%) and are not lower than 30% or above 140% [25].

Table 4. Recoveries and RSDs of the test analytes from the spiked chub homogenate.

Analyte	Recovery (RSD) [a] (%)				
	1 µg/kg	5 µg/kg	15 µg/kg	30 µg/kg	60 µg/kg
Hexachloro-1,3-butadiene	101 (6)	96 (4)	124 (5)	98 (6)	97 (6)
Pentachlorobenzene	98 (4)	88 (2)	103 (10)	103 (15)	90 (13)
alpha-HCH	99 (7)	86 (5)	101 (17)	91 (17)	62 (15)
Hexachlorobenzene	103 (5)	100 (4)	101 (16)	97 (6)	94 (6)
beta-HCH	95 (10)	82 (7)	86 (16)	92 (18)	58 (14)
Lindane	94 (10)	88 (5)	96 (16)	90 (18)	59 (15)
delta-HCH	91 (10)	85 (16)	89 (16)	88 (18)	57 (14)
Heptachlor	91 (7)	94 (8)	89 (12)	86 (9)	83 (12)
BDE-28	101 (6)	93 (10)	94 (18)	94 (3)	93 (14)
BDE-47	98 (6)	110 (2)	95 (6)	94 (3)	92 (15)
BDE-100	100 (5)	100 (5)	100 (14)	96 (5)	96 (10)
BDE-99	101 (9)	100 (8)	101 (15)	95 (12)	96 (8)
BDE-154	105 (4)	104 (12)	99 (11)	94 (6)	94 (7)
BDE-153	98 (11)	98 (12)	102 (9)	94 (12)	93 (5)

[a] $n = 5$.

3.3.4. Accuracy

The accuracy of the method was studied in terms of two components—precision and trueness [26]. The precision was evaluated as intra-day (PRE_{intra}) and inter-day (PRE_{inter}) precision and expressed by RSD for the repeated analyses of QC samples (lipid content of 5.2%) spiked with the test analytes at 5 µg/kg. The PRE_{intra} was determined by the analysis of ten replicates QC samples on one day, while PRE_{inter} was calculated from measurements of four replicates QC samples per day analyzed on four consecutive days. The trueness of the method was evaluated on the basis of ten measurements from the determination of PRE_{intra} and expressed as a mean recovery (R) and a mean relative bias (B_r). The results from the accuracy assessment are presented in Table 3. It can be seen that RSDs for PRE_{intra} and PRE_{inter} are in the ranges of 0.5%–11% and 3.2%–16%, respectively, showing a satisfactory precision of the method. The good trueness of the method is demonstrated by the values of Rec and B_r in the ranges of 87%–107% and −13.0%–7.4%, respectively.

Finally, the method's accuracy was studied by the analysis of the NIST SRM 1946 standard fish tissue reference material prepared from lake trout. Table 5 presents results obtained for those test analytes for which certified concentrations were available. According to obtained trueness and precision, acceptable results (trueness in the range 70%–120%, RSD ≤ 20%) were obtained for eight from nine analytes (except BDE-28). However, when comparing with the certified ranges, three results (lindane, BDE-28 and BDE-99) were outside and one result (BDE-154) was at the border of the certified range.

Table 5. Results from determination of selected chlorinated pesticides and PBDEs in the standard reference material NIST SRM 1946 (Lake Superior Fish Tissue).

Analyte	Certified Value [a] (µg/kg)	Determined Value [a] (µg/kg)	Trueness (RSD) (%)
Hexachlorobenzene	7.25 ± 0.83	6.47 ± 1.5	89 (2)
alpha-HCH	5.72 ± 0.65	5.44 ± 1.4	95 (7)
Lindane	1.14 ± 0.18	0.89 ± 0.26	78 (5)
BDE-28	0.742 ± 0.027	0.467 ± 0.067	63 (5)
BDE-47	29.9 ± 2.3	30.2 ± 5.1	101 (5)
BDE-99	18.5 ± 2.1	22.0 ± 3.7	119 (16)
BDE-100	8.57 ± 0.52	9.04 ± 1.8	105 (9)
BDE-153	2.81 ± 0.41	3.16 ± 0.69	112 (9)
BDE-154	5.77 ± 0.80	6.57 ± 1.2	114 (12)

[a] Mean value ± expanded combined measurement uncertainty ($U_{r,tot}$); $n = 3$.

3.3.5. Uncertainty of Measurement

To evaluate the uncertainty of measurement, a top-down approach has been used that utilizes data from validation and QC charts and is much simpler than the GUM (Generalized Uncertainty Method) bottom-up approach [27]. The combined measurement uncertainty was estimated according to Section 2.7 and the resulting values together with the individual uncertainty components are summarized in Table 6. As can be seen in Table 6, generally the most significant contribution to the measurement uncertainty was associated with the random error characterized by the within-lab reproducibility ($u_{r,repro}$). In several cases, the highest uncertainty component was the relative bias (B_r) representing the systematic error of the measurement. The B_r values were in the broadest range among the evaluated uncertainty components from 0.2 to −13.0%. The resulting values of the expanded combined uncertainty ($U_{r,tot}$) for the test analytes were in the range between 14.4 and 28.7%, being in accordance with the requirement (50%) of the EU guidance document SANTE/11813/2017 [25]. For comparison, in the study combining the QuEChERS method with GC–MS analysis for the determination of OC compounds in fish, Olivares et al. [28] estimated the combined measurement uncertainty for hexachlorobenzene and lindane at levels of 29.9 and 20.0%, respectively. In the work dealing with determination of OC pesticides in meat employing ASE, mini-silica column purification and GC–ECD analysis, Dimitrova et al. [29] obtained for hexachlorobenzene and HCH isomers expanded uncertainties in the range of 14.6%–17.9%. In the currently published study [30] concerning the determination of halogenated flame retardants by GC–API-MS/MS and GC–EI-MS after ASE and multi column clean-up, the values of expanded measurement uncertainties for the PBDEs of our interest in fish fillet were below 50%.

Table 6. Summary of uncertainties obtained for the test analytes using the top-down approach.

Analyte	$u_{r,repro}$ (%)	B_r (%)	$u_{r,cm}$ (%)	$u_{r,ref}$ (%)	$u_{r,tot}$ (%)	$U_{r,tot}$ (%)
Hexachloro-1,3-butadiene	8.01	−4.66	0.909	1.15	9.38	18.8
Pentachlorobenzene	9.90	−5.15	0.676	0.250	11.2	22.4
alpha-HCH	7.83	−10.6	0.154	0.475	13.2	26.3
Hexachlorobenzene	8.42	7.38	0.274	1.00	11.2	22.5
beta-HCH	8.13	−11.6	0.954	0.150	14.2	28.4
Lindane	6.04	−13.0	0.451	0.866	14.4	28.7
delta-HCH	7.15	−9.40	1.34	0.330	11.9	23.8
Heptachlor	7.60	−6.23	2.74	0.250	10.2	20.4
BDE-28	6.48	−1.28	2.78	0.295	7.18	14.4
BDE-47	7.98	1.90	1.56	0.300	8.36	16.7
BDE-100	9.75	0.192	0.913	0.300	9.80	19.6
BDE-99	8.01	−0.879	2.06	0.300	8.32	16.6
BDE-154	8.50	1.17	2.89	0.300	9.06	18.1
BDE-153	9.00	5.08	3.50	0.300	10.9	21.8

Abbreviations: $u_{r,repro}$—within-lab reproducibility; B_r—relative bias; $u_{r,cm}$—uncertainty of systematic error; $u_{r,ref}$—uncertainty of purity of analytical standard; $u_{r,tot}$—relative combined measurement uncertainty; $U_{r,tot}$—expanded combined measurement uncertainty.

3.4. Application of the Method to Real Samples

The applicability of the proposed method was evaluated by extraction and determination of the test analytes in homogenate samples of eight different fish species listed in Table 7. The lipid content of the analyzed fish (see Table 7) was in the range from 0.63% to 16% and the moisture content in the ranged from 58% to 81%, respectively. Table 7 presents the determined concentrations of the test analytes in fish homogenates and the relative recoveries (*RR*) of the analytes determined after their addition to the sample at concentration of 10 µg/kg. In general, the fish with the highest lipid content were the most contaminated (trout, salmon), while the least contaminated were those with the lowest lipid content (crucian carp, pollock). For all samples, hexachlorobenzene and BDE-47 were the most frequently detected analytes as well as the ones present at the highest concentrations. The *RR* values for the test analytes were in the range of 53%–128% with RSDs in the range of 1%–23%.

Table 7. Analysis of samples of different fish species.

Analyte	European chub Concentr. [a]/RR [b] (µg/kg/%)	Crucian carp Concentr. [a]/RR [b] (µg/kg/%)	European perch Concentr. [a]/RR [b] (µg/kg/%)	Northern pike Concentr. [a]/RR [b] (µg/kg/%)	Zander Concentr. [a]/RR [b] (µg/kg/%)	Brown trout Concentr. [a]/RR [b] (µg/kg/%)	Atlantic salmon Concentr. [a]/RR [b] (µg/kg/%)	Alaska pollock Concentr. [a]/RR [b] (µg/kg/%)
Hexachloro-1,3-butadiene	<0.09/96 (2)	<0.09/86 (6)	<0.09/87 (1)	<0.09/95 (1)	<0.09/93 (1)	<0.09/116 (1)	0.90 ± 0.01/98 (4)	0.22 ± 0.01/104 (3)
Pentachlorobenzene	<0.12/92 (3)	<0.12/96 (6)	<0.12/90 (2)	<0.12/105 (4)	<0.12/87 (3)	<0.12/108 (3)	0.22 ± 0.01/95 (1)	<0.12/104 (3)
alpha-HCH	<0.10/85 (2)	<0.10/98 (8)	<0.10/84 (2)	<0.10/92 (8)	<0.10/83 (6)	<0.10/105 (4)	0.23 ± 0.01/86 (3)	<0.10/95 (2)
Hexachlorobenzene	1.00 ± 0.01/95 (2)	0.35 ± 0.01/92 (3)	0.48 ± 0.01/93 (1)	1.84 ± 0.02/96 (3)	0.70 ± 0.01/95 (2)	1.12 ± 0.01/118 (2)	2.68 ± 0.05/96 (4)	0.18 ± 0.01/99 (1)
beta-HCH	<0.12/82 (3)	<0.12/93 (8)	<0.12/82 (2)	<0.12/97 (11)	<0.12/76 (7)	0.80 ± 0.02/113 (7)	0.12 ± 0.01/81 (4)	<0.12/93 (3)
Lindane	<0.13/83 (3)	<0.13/93 (8)	<0.13/82 (2)	<0.13/91 (9)	<0.13/79 (6)	<0.13/101 (6)	<0.13/84 (4)	<0.13/91 (2)
delta-HCH	<0.12/82 (2)	<0.12/92 (8)	<0.12/82 (3)	<0.12/102 (10)	<0.12/78 (7)	0.13 ± 0.003/100 (7)	<0.12/82 (4)	<0.12/90 (3)
Heptachlor	<0.13/78 (6)	<0.13/64 (12)	<0.13/98 (3)	<0.13/71 (6)	<0.13/114 (8)	<0.13/106 (8)	<0.13/83 (4)	<0.13/65 (10)
BDE-28	<0.12/89 (7)	<0.12/128 (9)	<0.12/88 (2)	<0.12/110 (5)	<0.12/99 (8)	<0.12/87 (4)	<0.12/86 (4)	<0.12/114 (9)
BDE-47	0.69 ± 0.01/95 (2)	<0.09/100 (2)	1.33 ± 0.04/103 (5)	0.21 ± 0.01/104 (6)	1.45 ± 0.01/103 (3)	0.36 ± 0.01/94 (4)	0.43 ± 0.02/97 (5)	<0.09/94 (0.4)
BDE-100	0.20 ± 0.002/100 (9)	<0.09/94 (8)	0.26 ± 0.02/111 (5)	<0.09/105 (9)	0.18 ± 0.01/96 (6)	0.10 ± 0.01/97 (6)	0.09 ± 0.01/105 (10)	<0.09/71 (12)
BDE-99	<0.07/96 (9)	<0.07/91 (8)	0.41 ± 0.02/113 (6)	<0.07/111 (14)	<0.07/96 (6)	0.25 ± 0.01/94 (5)	<0.07/104 (9)	<0.07/67 (4)
BDE-154	<0.16/103 (10)	<0.16/100 (21)	<0.16/120 (3)	<0.16/128 (15)	<0.16/99 (9)	<0.16/97 (7)	<0.16/112 (13)	<0.16/53 (15)
BDE-153	<0.14/99 (14)	<0.14/103 (23)	<0.14/119 (3)	<0.14/114 (19)	<0.14/95 (7)	<0.14/95 (6)	<0.14/100 (2)	<0.14/53 (16)
Lipid content (%)	3.5	0.96	3.0	2.4	1.5	8.2	16	0.63
Moisture content (%)	78	74	74	78	77	71	58	81

[a] For positive samples: mean value ± SD; n = 3. For negative samples: <LOQ value. [b] RR—relative recovery; in parenthesis: RSD value; n = 5.

In Figure 2, chromatograms from the analysis of low fat (perch, 3.0%) and high fat (salmon, 16%) fish samples are shown. In both total MRM chromatograms a high selectivity for the detected analytes and absence of matrix components can be observed. A difference between the counts obtained for ISs in the first and second chromatograms can be seen that can be related to the effect of the sample lipid content on the analysis. This is demonstrated in Figure 3, where a dependence of the IS peak size on the lipid content of the analyzed fish is presented. Therefore, the ISs and analytes peak areas generally decreased with the increasing lipid content of the matrix.

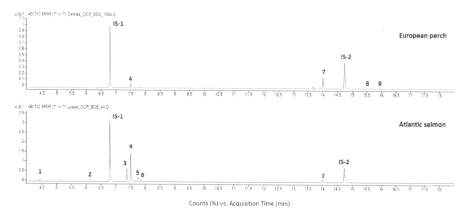

Figure 2.   Total ion MRM chromatograms from the GC–QqQ-MS/MS analysis of extracts prepared from samples of European perch and Atlantic salmon.   Peaks: 1—hexachloro-1,3-butadiene, 2—pentachlorobenzene, IS-1—tetrachloro-*m*-xylene, 3—*alpha*-HCH, 4—hexachlorobenzene, 5—beta-HCH, 6—lindane, 7—BDE-47, IS-2—BDE-77, 8—BDE-100, 9—BDE-99.

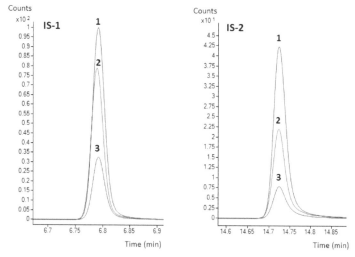

Figure 3. Dependence of the IS peak size on the lipid content of analyzed fish matrix. 1—perch (3.0%), 2—trout (8.2%), 3—salmon (16%). Other parameters of these fish samples are presented in Table 7.

3.5. Method's Analytical Eco-Scale Evaluation

The environmental impact of the proposed method was assessed using an analytical Eco-Scale approach that is based on assigning penalty points to parameters of the analytical process (depending on the use of hazardous chemicals, energy consumption, waste generation, etc.) that are not in agreement with the principles of green chemistry [31]. The analytical Eco-Scale analysis results in a score calculated by subtracting the penalty points from a value of 100, which represents an ideal green analysis. The assessment of the proposed method according to criteria set by Gałuszka et al. [31] is presented in Table 8. Due to obtained analytical Eco-Scale total score of 68, the current method can be classified as "an acceptable green analysis method" with low consumption of hazardous solvents.

Table 8. Analytical Eco-Scale assessment of the proposed method according to Galuszka et al. [31].

	Penalty Points
Reagents	
MeCN (5 mL)	4
$CHCl_3$ (50 µL)	2
Hexane (80 µL)	8
Analytes standard solution	4
H_2O (4 mL)	0
H_2SO_4 (1 mL)	2
$MgSO_4$ (2 g)	0
NaCl (0.5 g)	0
CH_3COONa	0
Instruments	
Vortex	1
Centrifuge	1
GC–MS/MS	3
Occupational hazard	3
Waste	4
Total penalty points	Σ 32
Analytical Eco-Scale total score	68

3.6. Comparison of the Proposed Method with Other Reported QuEChERS Based Methods

A comparison of the proposed method with other QuEChERS based methods with enhanced sample extract clean-up for determination of test analytes in fish is presented in Table 9. It can be seen that most of the applied clean-up procedures involve dual dSPE, which in several cases is combined with a freezing-out step for the low temperature lipid precipitation. The use of dSPE requires multiple weighing operations or purchase of custom-made sorbent blends and the freezing-out step significantly prolongs the sample preparation time. In the method used for determination of PBDEs (and other persistent organic pollutants) in salmon fillets [16], the ethyl acetate crude extract was purified applying gel permeation chromatography (GPC) and SPE. This clean-up procedure is rather laborious, requires a GPC instrument and is associated with high solvent consumption (ca. 150 mL per sample). The clean-up procedure described in the present paper is simple, fast, low cost, providing high co-extractives removal efficiency (involves complete removal of fatty acids), but it is only appropriate for the analysis of H_2SO_4 stable organic compounds. The LOQs of the presented method belong among the lowest listed in Table 9.

Table 9. Comparison of the developed method with other QuEChERS based methods with enhanced sample extract clean-up for determination of test analytes in fish.

Analytes	Extractant	Clean-Up	Analysis	Recoveries (%)	LOQs (µg/kg)	Reference
Pesticides	MeCN	Dual dSPE (1. PSA + C18 + MgSO$_4$; 2. PSA + C18 + MgSO$_4$)	GC–ECD	57–98	1.5–3.5	[9]
Pesticides	MeCN or MeCN/THF (3:1)	Freezing (2 h), dual dSPE (1. CaCl$_2$; 2. PSA + MgSO$_4$)	GC–MS	43–113	1–10	[12]
Pesticides	MeCN + CHCl$_3$ (10:1)	Dual dSPE (1. PSA + SAX +NH$_2$ + MgSO$_4$; 2. C18), freezing (overnight)	GC–MS	61–102	4–6	[13]
Pesticides	MeCN + hexane (15:2)	Freezing (20 min), dual dSPE (1. CaCl$_2$ + MgSO$_4$; 2. PSA + florisil + C18 +MgSO$_4$)	GC–MS/MS	60–127	2–13	[14]
PBDEs	MeCN (sonication)	Dual dSPE (1. PSA + C18 + MgSO$_4$; 2. PSA + C18 + MgSO$_4$)	GC–MS	60–107	<15	[15]
PBDEs	Ethyl acetate	GPC, SPE (silica + Na$_2$SO$_4$)	GC–MS	88–140	0.09–2.2	[16]
Pesticides	MeCN	Freezing (min. 4 h), dSPE (Z-Sep + MgSO$_4$), filtration (0.2 µm PTFE filter)	GC–MS/MS	86–101	0.08–0.15	[32]
PBDEs	MeCN + toluene (4:1)	Dual dSPE (1. EMR-Lipid; 2. Z-Sep + MgSO$_4$)	GC–MS/MS	79–116 (muscle) 89–107 (liver)	0.015–0.065 0.85–1.1	[33]
Pesticides, PBDEs	MeCN	pH-tuned DLLME (0.5 M CH$_3$COONa, CHCl$_3$), H$_2$SO$_4$ clean-up	GC–MS/MS	57–124 (pesticides) 93–110 (PBDEs)	0.09–0.17 0.07–0.16	This work

Abbreviations: MeCN—acetonitrile; dSPE—dispersive solid-phase extraction; PSA—primary secondary amine; C18—octadecyl silica; GC—gas chromatography; ECD— electron-capture detector; THF—tetrahydrofurane; MS—mass spectrometry; SAX—strong anion exchange resin; GPC—gel permeation chromatography; SPE—solid-phase extraction cartridge; Z-Sep; EMR-Lipid—clean-up sorbents; PTFE—polytetrafluoroethylene (Teflon); DLLME—dispersive liquid–liquid microextraction.

4. Conclusions

In this work, a rapid and non-laborious method was proposed for the determination of selected H_2SO_4 stable OC compounds and PBDEs in fish samples. The method employing QuEChERS sample preparation with pH-tuned DLLME and H_2SO_4 digestion fish extract clean-up followed by GC–QqQ-MS/MS analysis has successfully passed the validation process. The results obtained from the analysis of nine different fish species samples show the applicability of the method for the determination of selected analytes in fish. According to analytical Eco-Scale evaluation, the proposed method can be classified as "an acceptable green analysis method" with low consumption of hazardous solvents. The comparison of the method with other reported QuEChERS based methods shows its advantages such as simplicity, rapidity, low cost, high extract clean-up efficiency and good sensitivity.

Author Contributions: Conceptualization, P.T.; methodology, P.T. and S.N.; validation, S.N.; formal analysis, S.N.; investigation, S.N.; data curation, S.N. and P.T.; writing—original draft preparation, P.T. and S.N.; writing—review and editing, P.T.; supervision, P.T.

Funding: The authors acknowledge the financial support from the EU Cohesion Funds within the project Monitoring and assessment of water body status (No. 310011A366 Phase III).

Conflicts of Interest: The authors declare no conflict of interest.

References

1. Haggblom, M.M.; Bossert, I.D. Halogenated organic compounds: A global perspective. In *Dehalogenation: Microbial Processes and Environmental Applications*; Haggblom, M.M., Bossert, I.D., Eds.; Kluwer Academic Publishers: Norwell, MA, USA, 2003; pp. 3–32.
2. United Nations Environment Program (UNEP); Stockholm Convention on Persistent Organic Pollutants. *Adoption of Amendments of Annexes A, B and C*; United Nations Environment Program: Geneva, Switzerland, 2009.
3. European Commission. Regulation (EC) No 396/2005 of the European Parliament and of the Council of 23 February 2005 on maximum residue levels of pesticides in or on food and feed of plant and animal origin and amending Council Directive 91/414/EEC. *Offic. J. Eur. Commun.* **2005**, *L 70*, 1–16.
4. European Commission. 2014/118/EU: Commission Recommendation of 3 March 2014 on the monitoring of traces of brominated flame retardants in food. *Offic. J.* **2014**, *L 65*, 39–40.
5. European Commission. Decision No 2455/2001/EC of the European Parliament and of the Council of 20 November 2001 establishing the list of priority substances in the field of water policy and amending Directive 2000/60/EC. *Offic. J. Eur. Commun.* **2001**, *L 331*, 1–5.
6. US EPA. List of Priority Pollutants. Available online: http://water.epa.gov/scitech/methods/cwa/pollutants.cfm (accessed on 17 July 2018).
7. Chung, S.W.C.; Chen, B.L.S. Determination of organochlorine pesticide residues in fatty foods: A critical review on the analytical methods and their testing capabilities. *J. Chromatogr. A* **2011**, *1218*, 5555–5567. [CrossRef] [PubMed]
8. Berton, P.; Lana, N.B.; Ríos, J.M.; García-Reyes, J.F.; Altamirano, J.C. State of the art of environmentally friendly sample preparation approaches for determination of PBDEs and metabolites in environmental and biological samples: A critical review. *Anal. Chim. Acta* **2016**, *905*, 24–41. [CrossRef] [PubMed]
9. Tölgyessy, P.; Mihálikova, Z.; Matulová, M. Determination of selected chlorinated priority substances in fish using QuEChERS method with dual dSPE clean-up and gas chromatography. *Chromatographia* **2016**, *79*, 1561–1568. [CrossRef]
10. Pietroń, W.J.; Małagocki, P. Quantification of polybrominated diphenyl ethers (PBDEs) in food. A review. *Talanta* **2017**, *167*, 411–427. [CrossRef] [PubMed]
11. Rejczak, T.; Tuzimski, T. A review of recent developments and trends in the QuEChERS sample preparation approach. *Open Chem.* **2015**, *13*, 980–1010. [CrossRef]
12. Norli, H.R.; Christiansen, A.; Deribe, E. Application of QuEChERS method for extraction of selected persistent organic pollutants in fish tissue and analysis by gas chromatography mass spectrometry. *J. Chromatogr. A* **2011**, *1218*, 7234–7241. [CrossRef]

13. Molina-Ruiz, J.M.; Cieslik, E.; Cieslik, I.; Walkowska, I. Determination of pesticide residues in fish tissues by modified QuEChERS method and dual-d-SPE clean-up coupled to gas chromatography-mass spectrometry. *Environ. Sci. Pollut. Res. Int.* **2015**, *22*, 369–378. [CrossRef] [PubMed]

14. Chatterjee, N.S.; Utture, S.; Banerjee, K.; Ahammed Shabeer, T.P.; Kamble, N.; Mathew, S.; Ashok Kumar, K. Multiresidue analysis of multiclass pesticides and polyaromatic hydrocarbons in fatty fish by gas chromatography tandem mass spectrometry and evaluation of matrix effect. *Food Chem.* **2016**, *196*, 1–8. [CrossRef] [PubMed]

15. Morrison, S.A.; Sieve, K.K.; Ratajczak, R.E.; Bringolf, R.B.; Belden, J.B. Simultaneous extraction and cleanup of high-lipid organs from white sturgeon (Acipenser transmontanus) for multiple legacy and emerging organic contaminants using QuEChERS sample preparation. *Talanta* **2016**, *146*, 16–22. [CrossRef] [PubMed]

16. Cloutier, P.L.; Fortin, F.; Groleau, P.E.; Brousseau, P.; Fournier, M.; Desrosiers, M. QuEChERS extraction for multi-residue analysis of PCBs, PAHs, PBDEs and PCDD/Fs in biological samples. *Talanta* **2017**, *165*, 332–338. [CrossRef]

17. Han, L.; Matarrita, J.; Sapozhnikova, Y.; Lehotay, S.J. Evaluation of a recent product to remove lipids and other matrix co-extractives in the analysis of pesticide residues and environmental contaminants in foods. *J. Chromatogr. A* **2016**, *1449*, 17–29. [CrossRef] [PubMed]

18. Tölgyessy, P.; Nagyová, S. Rapid sample preparation method with high lipid removal efficiency for determination of sulphuric acid stable organic compounds in fish samples. *Food Anal. Methods* **2018**, *11*, 2485–2496. [CrossRef]

19. Tölgyessy, P.; Mihálikova, Z. Rapid determination of total lipids in fish samples employing extraction/partitioning with acetone/ethyl acetate solvent mixture and gravimetric quantification. *Food Control* **2016**, *60*, 44–49. [CrossRef]

20. Frenich, A.G.; Martínez Vidal, J.L.; Fernández Moreno, J.L.; Romero-González, R. Compensation for matrix effects in gas chromatography–tandem mass spectrometry using a single point standard addition. *J. Chromatogr. A* **2009**, *1216*, 4798–4808. [CrossRef] [PubMed]

21. Suchánek, M.; Friedecký, B.; Kratochvíla, J.; Budina, M.; Bartoš, V. Recommendations for the determination of uncertainties in the results of measurements/clinical tests in clinical laboratories. *Klin. Biochem. Metab.* **2006**, *14*, 43–53. (In Czech)

22. L'Homme, B.; Scholl, G.; Eppe, G.; Focant, J.F. Validation of a gas chromatography–triple quadrupole mass spectrometry method for confirmatory analysis of dioxins and dioxin-like polychlorobiphenyls in feed following new EU Regulation 709/2014. *J. Chromatogr. A* **2015**, *1376*, 149–158. [CrossRef]

23. Munaretto, J.S.; Ferronato, G.; Ribeiro, L.C.; Martins, M.L.; Adaime, M.B.; Zanella, R. Development of a multiresidue method for the determination of endocrine disrupters in fish fillet using gas chromatography–triple quadrupole tandem mass spectrometry. *Talanta* **2013**, *116*, 827–834. [CrossRef]

24. Sapozhnikova, Y.; Lehotay, S.J. Multi-class, multi-residue analysis of pesticides, polychlorinated biphenyls, polycyclic aromatic hydrocarbons, polybrominated diphenyl ethers and novel flame retardants in fish using fast, low-pressure gas chromatography–tandem mass spectrometry. *Anal. Chim. Acta* **2013**, *758*, 80–92. [CrossRef]

25. European Commission. Guidance Document on Analytical Quality Control and Method Validation Procedures for Pesticides Residues Analysis in Food and Feed, SANTE/11813/2017. Supersedes SANTE/11945/2015, Implemented by 01/01/2018. Available online: https://ec.europa.eu/food/sites/food/files/plant/docs/pesticides_mrl_guidelines_wrkdoc_2017-11813.pdf (accessed on 16 November 2018).

26. ISO 5725-1. *Accuracy (Trueness and Precision) of Measurement Methods and Results—Part 1: General Principles and Definitions*; International Organization for Standardization: Geneva, Switzerland, 1994.

27. ConsultGLP. Measurement Uncertainty—Comparing GUM and Top down Approaches. Available online: https://consultglp.com/2017/04/27/measurement-uncertainty-comparing-gum-and-top-down-approaches/ (accessed on 2 March 2019).

28. Olivares, I.R.B.; Costa, S.P.; Camargo, R.S.; Pacces, V.H.P. Development of a rapid and sensitive routine method of analyses for organochlorine compounds in fish: A metrological approach. *Pharm. Anal. Acta* **2016**, *7*, 502.

29. Dimitrova, R.T.; Stoykova, I.I.; Yankovska-Stefanova, T.T.; Yaneva, S.A.; Stoyanchev, T.T. Development of analytical method for determination of organochlorine pesticides residues in meat by GC-ECD. *Revue Méd. Vét.* **2018**, *169*, 77–86.

30. Neugebauer, F.; Dreyer, A.; Lohmann, N.; Koschorreck, J. Determination of halogenated flame retardants by GC-API-MS/MS and GC-EI-MS: A multi-compound multi-matrix method. *Anal. Bioanal. Chem.* **2018**, *410*, 1375–1387. [CrossRef] [PubMed]
31. Gałuszka, A.; Migaszewski, Z.M.; Konieczka, P.; Namieśnik, J. Analytical Eco-Scale for assessing the greenness of analytical procedures. *Trends Anal. Chem.* **2012**, *37*, 61–72. [CrossRef]
32. Baduel, C.; Mueller, J.F.; Tsai, H.; Gomez Ramos, M.J. Development of sample extraction and clean-up strategies for target and non-target analysis of environmental contaminants in biological matrices. *J. Chromatogr. A* **2015**, *1426*, 33–47. [CrossRef] [PubMed]
33. Cruz, R.; Marques, A.; Casal, S.; Cunha, S.C. Fast and environmental-friendly methods for the determination of polybrominated diphenyl ethers and their metabolites in fish tissues and feed. *Sci. Total Environ.* **2019**, *646*, 1503–1515. [CrossRef]

Article

Relationship between Water Activity and Moisture Content in Floral Honey

Chiachung Chen

Department of Bio-industrial Mechatronics Engineering, National Chung Hsing University, 250 Kuokuang Road, Taichung 40227, Taiwan; ccchen@dragon.nchu.edu.tw; Tel.: +886-4-22857562; Fax: +886-4-22857135

Received: 24 December 2018; Accepted: 11 January 2019; Published: 16 January 2019

Abstract: The water activity (Aw) and moisture content (MC) data of floral honey at five temperatures were determined using the Aw method and it was found that temperature significantly affected the Aw/MC data. The linear equation could be used to express the relationship between Aw and MC of honeys. The empirical regression equations between parameters and temperature were established. To evaluate the factors affecting the Aw/MC data, we used categorical tests of regression analysis to assess the effect of the correlation between Aw and MC of honey and examined the factors affecting the regression parameters. Six datasets from five countries were selected from the literature. The significance of the levels of qualitative categories was tested by *t*-test. The slope of the relationship between Aw and MC was affected by the state of honey (liquid and crystallized). The intercepts were significantly affected by honey type (flower or honeydew), harvesting year, geographical collection site, botanical source and other factors. The outliers in the datasets significantly affected the results. With modern regression analysis, useful information on the correlation between Aw and MC could be found. The results indicated that no universal linear equation for Aw and MC could be used. The Aw value could be used as a criterion for the honey industry; then, the MC of honey could be calculated by the specific linear equation between Aw and MC.

Keywords: honey; water activity; moisture content; regression; categorical testing

---

1. Introduction

The use of honey has a long history. The sweet food made by bees obtaining nectar from flowers is called flower honey. When bees obtain the sweet secretions of aphids or other insects, the product is called honeydew [1]. Honey is usually maintained in a liquid state. The other is called crystallized honey; many factors such as chemical composition, a degree of supersaturation, viscosity, a fructose/glucose ratio, moisture and dextrine content, water activity, micro-crystals and nucleation seeds presence, age, storage temperature and thermal history all influence its properties [2].

Fermentation is a problem for honey. The liquid mixture contains water, fructose and acid, so yeast could develop when the water content reaches a certain level [3]. The higher the water content, the greater the occurrence of fermentation and spoilage. The water content in called moisture content (MC) in the food industry. Rockland [4] determined that the amount of the free water is really a response to the development of yeast, and not MC. The amount of free water could be described as water activity (Aw). The Aw for honey ranges from 0.5~0.65 [5–11]. The limiting Aw for yeast in honey is about 0.61~0.62 [12] according to Beuchat et al. [12] and 0.60 [8,10,13,14] according to Beckh et al. [13], Gletier et al. [8], Ruegg and Blanc [14], and Zamora and Chirife [10].

The official method for MC measurement is refractometric measurement. It is inexpensive and easy to use. However, it cannot be directly used for crystallized honey.

With a promising methods for measuring Aw, the MC of honey can then be calculated using a previously established empirical equation. Some experiments have been performed to determine the Aw and MC data of honey, subsequently establishing the relationship between Aw and MC by regression analysis. Several models have been proposed [5–7,9,11,13–19].

The factors affecting the Aw of honey included the type (flower or honey dew), state (liquid or crystallized), climatic and botanical origins, and induced fine granulation [5,6,10,19].

Glitter et al. [8] compared two types of honey: flower and honeydew, in liquid or crystallized state. At the same MC, Aw was higher for crystallized than for liquid honey. At the same Aw in liquid state, MC was higher for honeydew than for flower honey; there was no difference between the two types in the crystallized state. However, only correlation coefficients (r) between Aw and MC were reported in this study. No statistical methods were reported for comparison.

Cavia et al. [6] compared three groups of honey from different climatic regions and harvest years. The *t-test*, used to assess differences between groups, revealed no significant difference among the groups. The author then pooled the three groups of datasets and proposed an empirical equation. In comparing this equation with that in the literature, the intercepts and slopes of these equations differed. The authors mention these differences related to the measurement of Aw and MC.

Chirife et al. [7] established a linear equation for Aw and MC from 36 liquid Argentinian honey samples and compared this equation with that from Beck et al. [13]. The slope and intercept differed, and the results were attributed to the lack of accuracy of measurement techniques and the inference of different botanical sources and geographical collection sites. Abramovic et al. [5] investigated the correlation between Aw and MC for flower and honeydew honey in Slovenia. At the same MC, Aw was higher for honeydew than for flower honeys. The two datasets were pooled and a linear equation for Aw and MC was established. Comparing regression parameters with five other empirical equations in the literature, the difference found was attributed to the sugar composition and measurement of Aw. Zamora et al. [11] examined the data from Beckh et al. [13] in fluid, partially crystalline and crystalline honey and found no significant differences by F-test. However, the determination of coefficient R^2 was only 0.53 for the pooled data. Zamora et al. [11] found no significant difference in the relationship between Aw and MC for honey by botanical source or collection site. Comparing four regression equations in the literature, the authors found that the slope for the three datasets was similar [11]. Perez et al. [8] reviewed 10 datasets for Aw and MC in flower honey and proposed a weighted average regression equation by using meta-analysis.

Recently, modern regression analysis was proposed to analyze treatment [20–22]. If the influencing factor for the treatment has several levels of qualitative categories, the significance of these treatments can be tested by categorical testing with regression analysis [20–22].

The objectives of this study were to: (1) determine the Aw/MC relationships for floral honey between 10 and 30 °C; and (2) evaluate the factors affecting the sorption isotherms of honey with previously published data by using regression analysis.

2. Materials and Methods

2.1. Materials

The floral honey used for this study was Longyuan honey collected at the Chunglaun Township, Nantou, Taiwan. The initial moisture content of the sample was 17.16% on a w.b.

The desired moisture content for storage and processing ranges from 17% to 22% w.b. The samples were rewetted by adding the amount of the water necessary to reach the desired moisture content. The sample preparation was performed according to the study by Shen and Chen [23]. All samples were sealed in glass vessels and stored at 5 °C for three weeks to ensure uniform moisture content. The moisture content of the honey was measured by using the ATAGO DR-A1 ABBE Refractometer (Atago Inc., Bellevue, WA, USA). The total number of sample in this study was 27. There were nine moisture contents, and three replicates of each moisture content.

2.2. RH Meter

The THT-V2 humidity transmitter (Shinyei technology, Kobe, Japan) was used to determine the water activity of the floral honey. These RH sensors were calibrated with several saturated salt solutions, and the accuracy of the RH meter was within 0.7% RH after calibration. All measured RH values were transformed into actual values by using previously established calibration equations to enhance its accuracy.

2.3. Aw Method

The set-up for the Aw method is shown in Figure 1. Samples at known moisture contents were placed in a glass vessel, sealed with a rubber stopper to ensure airtight conditions, and then placed in a temperature-controlled chamber that was maintained at 5 °C. The volume of the vessel was 250 mL. When the temperature and the RH within the container were stabilized, the vapor pressure of the samples and the interstitial air in the vessel reached the equilibrium state. The RH and temperature values were determined. To ensure the equilibrium state, each temperature level was maintained for 12 h, then adjusted to next temperature level. All Aw values were measured at five temperatures (10, 15, 20, 25 and 30 °C). After finishing the experiments, the samples were taken from each vessel to determine the moisture content using the ATAGO DR-A1 ABBE Refractometer.

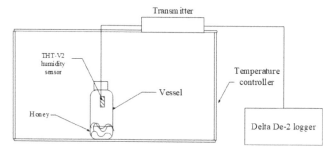

Figure 1. Diagram of the experimental set-up.

This technique has been used to determinate sorption isotherm for autoclaved aerated concrete [24], sweet potato slices [25], pea seeds [26] and Oolong tea [27].

2.4. Literature Survey

The six datasets from five countries used to evaluate the factors affecting the regression parameters between Aw and MC are in Table 1. The published models used, along with the data from the literature and seven other published models, are displayed in Table 2.

Table 1. Selected studies on the relationship between water activity and moisture content in honey.

Types	Geographical Original of Honeys	Aw Determination Method	Sample Size	Moisture Range (%)	Reference
Honeydew and flower	Slovenia	Cx-2T Chill-mirror Aw system	150	13.4-18.6	Abramovic et al. [5]
Flower	Spain (liquid and crystallized)	Cx-2T Chill-mirror Aw meter	90	14.2-21.5	Cavia et al. [6]
Flower	Argentine	Aqual series 3 Model TE dew-point Aw meter	35	13.8-20.8	Chirife et al. [7]

Table 1. *Cont.*

Types	Geographical Original of Honeys	Aw Determination Method	Sample Size	Moisture Range (%)	Reference
Honeydew and flower	Germany	Navasian Aw meter Aqual series 3	166	14.2-22.7	Gleiter et al. [8]
Flower	Spain	Model TE dew-point Aw meter	13	16.5~19.4	Sanjuan et al. [9]
Flower	Argentine (liquid and crystallized)	Aqual series 3 Model TE dew-point Aw meter	36	15.8~27.1	Zamora et al. [11]

Table 2. Published equations of water activity and moisture content of honey.

Study	Equations	R^2 (r)	Reference
	I. Datasets used in this study.		
1.	Aw = 0.23 + 0.019MC	(0.843)	Abramovic et al. [5]
2.	Aw = 0.2674 + 0.01955MC	(0.709)	Cavia et al. [6]
3.	Aw = 0.262 + 0.0179MC	0.969	Chirife et al. [7]
4.	Aw = 0.35732 + 0.01349MC	0.654	Sanjuan et al. [9]
5.	Aw = 0.305 + 0.0155MC	0.969	Zamora et al. [11]
	II. Datasets not used in this study.		
1.	Aw = 0.13 + 0.025MC	(0.8230)	Alcala and Gomez [15]
2.	Aw = 0.342 + 0.014MC	(0.723)	Beckh et al. [13]
3.	Aw = 0.25643 + 0.01965MC	(0.813)	Estupinan et al. [16]
4.	Aw = 0.38242 + 0.01211MC	(0.765)	Millan et al. [17]
5.	Aw = 0.2686 + 0.01756MC	Meta-analysis	Perez et al. [18]
6.	Aw = 0.0.271 + 0.0177MC	(0.901)	Ruegg and Blanc [14]
7.	Aw = 0.248 + 0.0175MC	(0.973)	Salamanca et al. [19]

2.5. Categorical Tests

If the influencing factor has several levels of qualitative categories, the significance of the qualitative treatment could be tested by *t*-test or F-test.

2.5.1. Testing the Slope for Two Treatments

To evaluate the effect of categorical variables such as type or state of honey, an indicator variable is used. The equation for the regression line relating two types of datasets that differ in both intercept and slope are as follows:

$$y = b_o + b_1 X_1 + b_2 Z_1 + b_3 X_1 Z_1 + \varepsilon \tag{1}$$

$$Z_1 = 0, if\ the\ data\ is\ from\ factor\ A,$$

$$Z_1 = 1, if\ the\ data\ is\ from\ factor\ B$$

$$For\ factor\ A: y = b_o + b_1 X_1 + \varepsilon \tag{2}$$

$$For\ factor\ A: y = (b_o + b_2) + (b_1 + b_3) X_1 + \varepsilon \tag{3}$$

$$H_0: b_2 = b_3 = 0$$

$$H_1: b_2 \neq 0, b_3 \neq 0$$

To test the hypothesis that two regression lines have the same slope or intercept, we could use the *t*-test.

2.5.2. Testing the Slope for Three Treatments

For three treatments, the regression equations relating datasets that differ in both intercept and slope are as follows:

$$y = b_0 + b_1 X_1 + b_2 Z_1 + b_3 Z_2 + b_4 X_1 Z_1 + b_5 X_1 Z_2 + \varepsilon \tag{4}$$

$$Z_1 = Z_2 = 0, if\ the\ observation\ is\ from\ A,$$

$$Z_1 = 1\ and\ Z_2 = 0, if\ the\ observation\ is\ from\ B,$$

$$Z_1 = 0\ and\ Z_2 = 1, if\ the\ observation is\ from\ C,$$

$$\text{For factor A}: \ y = b_0 + b_1 X_1 + \varepsilon \tag{5}$$

$$\text{For factor B}: \ y = (b_0 + b_2) + (b_1 + b_4) X_1 + \varepsilon \tag{6}$$

$$\text{For factor C}: \ y = (b_0 + b_3) + (b_1 + b_5) X_1 + \varepsilon \tag{7}$$

$$H_0 : b_4 = b_5 = 0$$

$$H_1 : b_4 \neq 0, b_5 \neq 0$$

2.5.3. Two Indicator Variables

If the qualitative variables have two qualitative factors (e.g., flower and honeydew, crystallized and liquefied), the regression line can be expressed as follows:

$$y = b_0 + b_1 X_1 + b_2 Z_1 + b_3 Z_2 + b_4 X_1 + b_5 X_1 Z_2 + \varepsilon \tag{8}$$

$$\text{a}: \text{ liquefied, flower}: \ Z_1 = 0, Z_2 = 0$$

$$y = b_0 + b_1 X_1 + \varepsilon \tag{9}$$

$$\text{b}: \text{ crystallized, flower}: \ Z_1 = 0, Z_2 = 1$$

$$y = (b_0 + b_3) + (b_0 + b_5) X_1 + \varepsilon \tag{10}$$

$$\text{c}: \text{ liquefied, honeydew}: \ Z_1 = 1, Z_2 = 0$$

$$y = (b_0 + b_2) + (b_0 + b_4) X_1 + \varepsilon \tag{11}$$

$$\text{d}: \text{ crystallized, honeydew}: \ Z_1 = 1, Z_2 = 1$$

$$y = (b_0 + b_2 + b_3) + (b_1 + b_4 + b_5) X_1 + \varepsilon \tag{12}$$

3. Results

3.1. Water Activity of Honey

The Aw/MC data at five temperatures are shown in Figure 2. Temperature significantly affected the Aw/MC data.

The results of the estimated parameters and comparison statistics for the linear equation at different temperature are in Table 3. The effect of temperature on parameters A and B is shown in Figures 3 and 4.

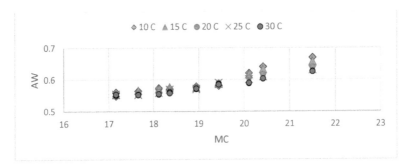

Figure 2. Effect of temperature on the water activity (Aw)/moisture content (MC) data of floral honey.

Table 3. Estimated values of parameters in the linear equation.

Temperature °C	Parameters		Coefficients of Determination R^2	Standard of Deviations of Estimated Values s
	A	B		
10	0.09520	0.026172	0.908	0.3195
15	0.14395	0.023511	0.936	0.2611
20	0.20243	0.020089	0.987	0.2165
25	0.22204	0.018839	0.968	0.1710
30	0.23771	0.017768	0.958	0.1536

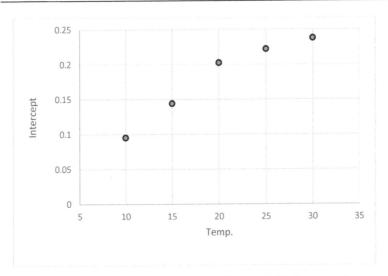

Figure 3. Effect of temperature on parameter A (intercept) of the linear equation.

The empirical regression equations between parameters and temperature were established. The equation for A and B was expressed as:

$$A = -0.06999 - 0.0019264\text{Temp} + 3.000030 \times 10^{-4}\text{Temp}^2, R^2 = 0.991 \tag{13}$$

$$B = 0.03516 - 0.0010345\text{Temp} + 1.4149 \times 10^{-5}\text{Temp}^2, R^2 = 0.990 \tag{14}$$

Three forms of the linear equation that incorporated the temperature term were proposed as follows:

$$Aw = (-0.06999 - 0.0019264\text{Temp} + 3.000030 \times 10^{-4}\text{Temp}^2 + 0.3516$$
$$-0.0010345\text{Temp} + 1.4149 \times 10^{-5}\text{Temp}^2)MC \tag{15}$$

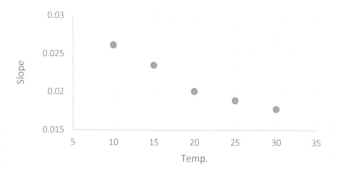

Figure 4. Effect of temperature on parameter B (slope) of the linear equation.

3.2. Comparison with Published Data

The Aw/MC linear equation of floral honey at 25 °C in this study was compared with published data (Figure 5). At MC < 20.5%, the Aw values of this study were lower than other data. However, when MC > 20.5%, the Aw values of this study were higher than those of Gleiter et al. [8]. The reason for this could be that the Aw/MC data were affected by honey type (flower or honeydew), harvesting year, geographical collection site, botanical source and other factors. Further study was executed to study the factors influencing the Aw/MC data.

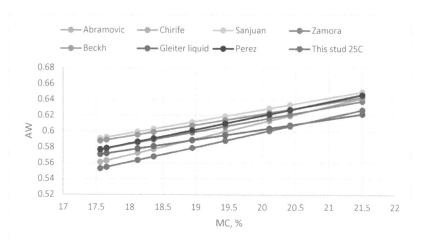

Figure 5. Comparison of Aw/MC equation at 25 °C with models from literature: [5,7–10,12,18].

3.3. Effect of Honey Type on Aw

Two types of honey (flower and honeydew) [5] were used to evaluate the factors affecting the relationship between Aw and MC by Equation (1). The data distribution and predicted lines are in Figure 6.

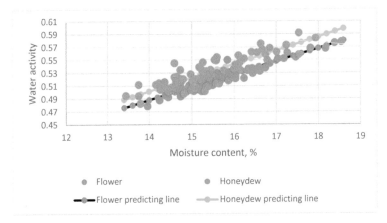

Figure 6. The relationship between water activity (Aw) and moisture content (MC) of flower and honeydew honey in Slovenia [5].

The results of the linear regression are as follows:

$$Aw = \underset{(11.45)}{0.20801} + \underset{(17.29)}{0.019985MC} - \underset{(-0.074)}{0.00189Z} + \underset{(0.66)}{0.00108Z{\cdot}MC} \tag{16}$$

$$R^2 = 0.823$$

where Z is the categorized variable, Z = 0 is flower honey and Z = 1 is honeydew honey. The numbers in parentheses below the estimated values of parameters are *t*-test values for the estimated value. The *t* and *p* values for Z·MC were 0.665 and 0.112, respectively. Type had no significant effect on the Aw and MC relationship.

The adequate equation was as follows:

$$Aw = \underset{(15.49)}{0.19964} + \underset{(25.06)}{0.020579MC} + \underset{(9.18)}{0.014766Z} \tag{17}$$

For flower honey,

$$Aw = 0.19964 + 0.020579MC \tag{18}$$

For honeydew honey,

$$Aw = 0.18171 + 0.020579MC \tag{19}$$

From Equation (17)–(19), we found no significant difference in the slope of the linear equation for flower and honeydew honey. However, the intercept significantly differed with two types of honey.

3.4. Effect of the Type and State of Honey on the Aw and MC Relationship

The datasets for Glitter et al. [8] included different honey types (flower and honeydew) and states (liquid and crystallized). Two indicator variables were analyzed by Equation (8).

The regression equation was as follows:

$$Aw = \underset{(24.12)}{0.30845} + \underset{(21.83)}{0.016905MC} + \underset{(14.76)}{0.018156Z_1} + \underset{(2.71)}{0.034193Z_2} - \underset{(-3.93)}{0.00391Z_2{\cdot}MC} \tag{20}$$

$$R^2 = 0.831$$

For crystallized flower honey, $Z_1 = 0$, $Z_2 = 0$,

$$Aw = 0.340845 + 0.016905MC \tag{21}$$

For liquid flower honey, $Z_1 = 0$, $Z_2 = 1.0$

$$Aw = 0.34264 + 0.012995MC \tag{22}$$

For crystallized honeydew honey, $Z_1 = 1.0$, $Z_2 = 0$

$$Aw = 0.32661 + 0.016905MC \tag{23}$$

For liquid honeydew honey, $Z_1 = 1.0$, $Z_2 = 1.0$

$$Aw = 0.36080 + 0.012995MC \tag{24}$$

The results indicated a significant difference in the intercept. With the same crystallized state, flower and honeydew honey had a similar slope, 0.016905. With the same liquid state, the slope was 0.012995. That is, the state not the type of honey significantly affects the slope parameter of the Aw linear equation. The prediction lines of two states and two types of honey are in Figure 7.

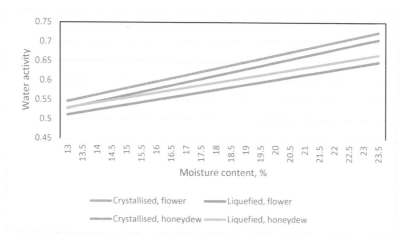

Figure 7. The prediction equations between water activity and moisture content including different type (flower and honeydew) and state (liquid and crystallized) of honey [8].

3.5. Comparison of the Correlation between Aw and MC with Two Datasets

Several datasets for Aw and MC for honey were used to evaluate factors affecting correlation between the Aw and MC.

3.5.1. Argentinian [7] and Slovenian Honeys [5]

The datasets from different countries with the liquid state are in Figure 8.

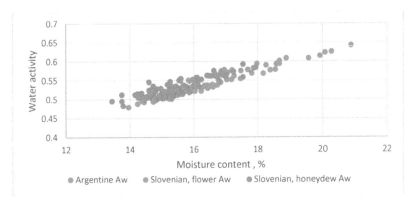

Figure 8. The relationship between water activity and moisture content of flower and honeydew honey from Argentinian [7] and Slovenia [5].

Two datasets for Slovenia honey were pooled and evaluated by Equation (4). The regression equation was as follows:

$$Aw = \begin{array}{cccc} 0.22826+ & 0.019149MC+ & 0.0412Z- & 0.00147Z{\cdot}MC \\ (15.64) & (20.39) & (1.449) & (-0.901) \end{array}$$

$$(25)$$

$$R^2 = 0.872$$

The *t*-test value for Z·MC was −0.901 and not significant. The adequate linear equation was as follows:

$$Aw = \begin{array}{ccc} 0.23581+ & 0.018662MC+ & 0.015256Z \\ (19.75) & (24.30) & (5.66) \end{array}$$

$$(26)$$

$$R^2 = 0.873$$

3.5.2. German and Slovenian Honeys

The datasets for Aw and MC for the two countries had the same slope, but a different intercept.
The datasets from Germany (pooled flower and honeydew, liquid state) [8] and Slovenia (liquid state) honey [5] are in Figure 9 and were used for assessing the influencing factors.

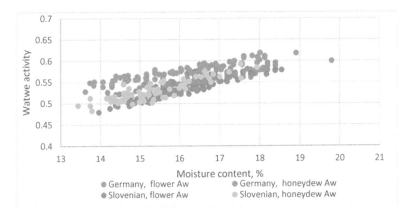

Figure 9. The relationship between water activity and moisture content for datasets for honey from Germany (pooled of the flower and honeydew, liquid state) [8] and Slovenia (liquid state) [5].

The linear equation was as follows:

$$Aw = 0.22826 + 0.019149MC + 0.095601Z - 0.00302Z \cdot MC$$
$$(14.40) \quad\quad (18.76) \quad\quad (4.88) \quad\quad (-1.080)$$

$$(27)$$

$$R^2 = 0.895$$

The *t*-test value for $Z \cdot MC$ was insignificant. The adequate equation is as follows:

$$Aw = 0.31305 + 0.013680MC + 0.026313Z$$
$$(28.94) \quad\quad (19.74) \quad\quad (15.31)$$

$$(28)$$

$$R^2 = 0.893$$

The type (flower or honeydew) did not significantly affect the slope.

3.5.3. Mixed-Source and Slovenian Honeys

The datasets from different types and states [13] and Slovenian honey (pooled liquid states: flower and honeydew) [5] are evaluate in Figure 10.

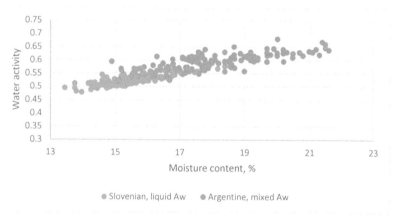

● Slovenian, liquid Aw   ● Argentine, mixed Aw

Figure 10. The relationship between water activity and moisture content for datasets for honey of different types and states [13] and Slovenia (pooled data of liquid states: flower and honeydew) [5].

The linear regression was as follows:

$$Aw = 0.22826 + 0.019149M + 0.11401Z - 0.00507Z \cdot MC$$
$$(13.31) \quad\quad (13.43) \quad\quad (4.0513) \quad\quad (-2.96)$$

$$(29)$$

$$R^2 = 0.861$$

The $Z \cdot MC$ had a *t*-test value of -2.96 and $p = 0.00336$, showing a significant effect. Both datasets had different slopes and intercepts.

Chirife et al. [7] compared the correlation for Argentina fluid honey and mixed honey from different countries [13] and found that the intercept and slopes of both differed. These results could be explained by the source of the honey. The Argentinian honey was liquid, the mixed honey included liquid, crystallized and partially crystallized states.

3.5.4. Spanish and Slovenian Honeys

The datasets from Spain (flower honey, unknown state) [6] and Slovenia (pooled data of liquid states, flower and honeydew) [5] are in Figure 11. The regression equation was as follows:

$$\text{Aw} = \underset{(13.41)}{0.22826+} \underset{(17.48)}{0.019149\text{MC}+} \underset{(5.03)}{0.094645\text{Z}-} \underset{(-2.30)}{0.00273\text{Z}\cdot\text{MC}} \tag{30}$$

$$R^2 = 0.899$$

The *t*-test value for Z·MC was −2.30, with *p* = 0.022, so datasets for the two countries had different slopes and intercepts.

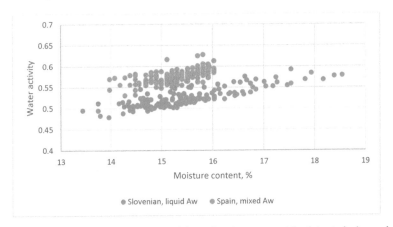

Figure 11. The relationship between water activity and moisture content for datasets for honey from Spain (flower honeys, unknown state) [6] and Slovenia (pooled data of liquid states, flower and honeydew) [5].

3.5.5. German (Crystallized State) and Slovenian (Liquid State) Honeys

The datasets from Gleiter et al. [8] and Abramovic et al. [5] for honey in different states are assessed (Figure 12).

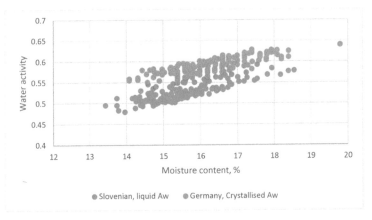

Figure 12. The relationship between water activity and moisture content for datasets for honey from Germany [8] and Slovenia [5].

The regression equation was as follows:

$$Aw = \underset{(12.01)}{0.22826+} \underset{(15.66)}{0.019149MC+} \underset{(6.46)}{0.14899Z-} \underset{(-5.34)}{0.00719Z \cdot MC} \tag{31}$$

$$R^2 = 0.746$$

The linear equations for the two datasets had different slopes and intercepts.

3.5.6. Comparing the Correlation between Aw and MC with Three Datasets

a. Case 1

Three datasets were used: German liquid flower ($Z_1 = 0$, $Z_2 = 0$) and honeydew ($Z_1 = 1$, $Z_2 = 0$) honey [8] and Argentinian liquid honey ($Z_1 = 0$, $Z_2 = 1$) [7].
The linear equation was as follows:

$$Aw = \underset{(39.61)}{0.31565+} \underset{(30.26)}{0.014452MC+} \underset{(4.87)}{0.009275Z_1+} \underset{(20.06)}{0.02677Z_2} \tag{32}$$

$$R^2 = 0.834$$

The slope of the three datasets was identical, and the intercepts significantly differed.

b. Case 2

Three datasets, Slovenian [5] liquid honeydew ($Z_1 = 0$, $Z_2 = 0$) and flower honey ($Z_1 = 1$, $Z_2 = 0$) and Argentinian liquid honey [7], were used.
The linear equation was as follows:

$$Aw = \underset{(21.62)}{0.21471+} \underset{(31.07)}{0.019558MC+} \underset{(9.15)}{0.020636Z_1+} \underset{(9.54)}{0.014421Z_2} \tag{33}$$

$$R^2 = 0.917$$

The slope of the three datasets was identical and the intercepts significantly differed.
Zamora et al. [10] compared regression equations for Aw and MC for honey from different countries and found no significant difference based on botanical source or geographical collection site. Our study confirms these results.

c. Case 3

Three datasets, German crystallized flower and honeydew honey ($Z_1 = 0$, $Z_2 = 0$) [8], Argentinian liquid honey ($Z_1 = 1$, $Z_2 = 0$) [7] and Slovenian liquid honeydew and flower honey ($Z_1 = 0$, $Z_2 = 1$) [5], were used. The linear equation was as follows:

$$Aw = \underset{(29.95)}{0.31724+} \underset{(14.77)}{0.01136MC-} \underset{(-6.70)}{0.14899Z_1-} \underset{(-3.38)}{0.10886Z_2+} \underset{(5.53)}{0.007789Z_1 \cdot MC+} \underset{(3.42)}{0.006316Z_2 \cdot MC} \tag{34}$$

$$R^2 = 0.782$$

The slopes and intercepts of the three datasets significantly differed.

3.5.7. Outlier Detection

The intercept and slope of the equation for the dataset of flower honey from La Palma Island, Spain, significantly differed from those in other datasets [9]. Outlier data (17.2233, 0.6084) was found by the Cook's distance test [21]. The original linear equation proposed by the authors was as follows:

$$Aw = 0.35732 + 0.01349MC, \; R^2 = 0.63 \tag{35}$$

After deleting this data, the new equation was as follows:

$$Aw = 0.32842 + 0.01495MC, \; R^2 = 0.93 \tag{36}$$

The comparison between Equations (32) and (33) is shown in Figure 13.

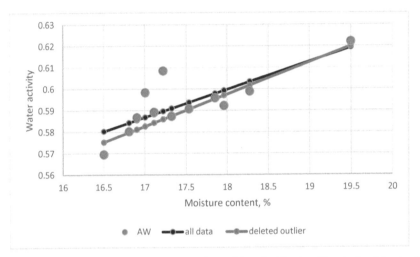

Figure 13. The comparison between two equations with and without outliers in the datasets of Sanjuan et al. [9].

After deleting this data point, the slope, intercept and coefficient of determination changed obviously.

If we compare all data from Sanjuan et al. [9] with the datasets for Argentinian honey [7], the linear equation was as follows:

$$Aw = \begin{array}{cccc} 0.26838+ & 0.017675MC- & 11.4662Z+ & 48.35758Z \cdot MC \\ (0.48) & (0.59) & (-3.22) & (8.15) \end{array} \tag{37}$$

$$R^2 = 0.960$$

The intercept and slope for the two datasets differed significantly.

If the outlier was deleted from the datasets of Sanjuan et al. [9], the equation for evaluating the two datasets was as follows:

$$Aw = \begin{array}{ccc} 0.27478+ & 0.017310MC+ & 0.013714Z \\ (25.40) & (28.19) & (7.31) \end{array} \tag{38}$$

$$R^2 = 0.956$$

The slope of the two datasets was identical. From the results of Equations (37) and (38), the outlier significantly affected the comparison results for the two datasets. With modern regression analysis, more useful information on correlation between Aw and MC could be found. The results indicated the importance of finding the correct equation with modern regression.

4. Discussion

Based on the study of the datasets of Gleiter et al. [8], the intercept parameters differed significantly. The slope parameter could be classified into two categories: liquid and crystallized. Thus, the slope for the linear equation was affected only by the state of the honey. The type of honey, flower and honeydew, and other factors did not affect the slope but did affect the intercept.

Chirife et al. [7] found that Aw in honey was determined mainly by the concentrations of fructose and glucose that are most abundant in honey. The authors developed an Aw equation from the effect of the osmotic concertation on the osmotic coefficient, which was as follows:

$$Aw = Exp(-\Phi 0.018mv) \tag{39}$$

where Φ is the osmotic coefficient, m is molality and v is the number of moles of kinetic units.

By Taylor's expansion, and assuming $0.018mv \ll 1$, the new relationship is as follows:

$$Aw = 1 - Km \tag{40}$$

For very concentrated and small intervals of sugar solutions, Equation (40) was rewritten as follows:

$$Aw = A - B(s) = b_0 + b_1 MC \tag{41}$$

where (s) is the solid concentration in water, and A and B are constants.

In this study, we found the slope to be affected only by the state of the honey (liquid or crystallized). The other factors, such as type (flower or honeydew), geographical collection sites and botanical source did not significantly affect the slope, but did affect the intercept of the Aw equation.

Perez et al. [18] selected 10 datasets of flower honey to study the relationship between Aw and MC and found similar but not identical the slopes and intercepts of these linear regressions. The authors attributed the finding to sampling error, accuracy of the Aw measurement, and variation in sugar composition. The slopes for 10 datasets ranged from 0.0149 to 0.0197. However, the state of honey was not mentioned in this research. We found a slope of 0.016905 for crystallized honey and 0.012995 for liquid honey for the datasets of Gleiter et al. [8]. The wide slope range for the 10 datasets from the study of Perez et al. [18] may be explained by the effects due to the state of honey. The slope for the linear equation for Slovenia honey [5] and five other datasets ranged from 0.014 to 0.0196. The difference in parameters was attributed to sugar composition and the Aw determination methods by the authors. The significant difference between the two maximum and minimum slope values, 0.014 and 0.0196, could be explained by the state of the honey.

The study by Cavia et al. [6] included three groups of samples. The G_1 and G_2 datasets were obtained in 1996 and 1998 from a continental climate, and the G_3 datasets was obtained in 1998 from an oceanic climate. The slopes for the Aw model were 0.02149 and 0.02362 for G_1 and G_2 and 0.01476 for G_3. The significant difference between the three slopes could be explained by the effect of climate on the state of the honey. The crystallized state enhanced by the continental climate may be due to the difference in slope values.

The MC of honey is considered to be the criterion for the honey industry. The MC is usually determined by the refractometric technique. The method is simple and inexpensive. However, the MC could be affected by weather conditions, original moisture content of the nectar, and environmental temperature and humidity after harvesting. The storage materials and sealed technique also affect MC. The MC of the crystallized state cannot be directly measured by refractometer.

The Aw is measured by some commercial equipment. The criterion of Aw < 0.6 may be used as a safety standard to prevent the development of osmotolerant yeasts. Recently, the performance of electronic hygrometers has been improved. They have been used to determine the Aw of tea leaves and other materials [23,24].

In this study, we found an effect of factors on slope for the correlation between Aw and MC. The state of honey, crystallized and liquid, had a significant effect on the slope value. However, other factors, such as harvesting year, botanic source and collection sites did not affect the slope but did affect the intercept. Therefore, no universal linear equation for Aw and MC could be established. The Aw value may be used as the criterion for the honey industry and directly determined by an electronic hygrometer. Then the MC of honey could be calculated by the specific linear equation between Aw and MC. The effect of the temperature needs to be considered.

In the traditional MC and AW determination method, honey must be liquefied previously, so that all crystals are totally melted, such that all measurements can be done with liquid honey. By the Aw method used in this study, the Aw values of crystallized honey could be determined directly in the crystalline state.

In this study, the moisture content of liquid honey was measured by using a refractometer. There are two official procedures Association Official Analytical Chemists (AOAC) and European Honey Commission (EHC) [28,29] for determining the moisture content. A comparison between the official method and the refractometer has been reported [30]. The comparison between the official method and the refractometer of floral honey used in this study will be further studied.

5. Conclusions

Conclusions were drawn from the results of this study.

The Aw/MC data at five temperatures were determined, and temperature significantly affected the Aw/MC data. The linear equation could be used to express the relationship between Aw and MC of Honeys. The empirical regression equations between parameters and temperature were established. The intercept and slope of the linear equation could be expressed as the polynomial equations.

The slope of the correlation between Aw and MC was affected by the state of honey (liquid and crystallized). The intercept was significantly affected by honey type (flower or honeydew), harvesting year, geographical collection site and botanical source. The outliers in the dataset significantly affected the comparison results. Modern regression analysis can provide useful information for the correlation between Aw and MC. No universal linear equation for Aw and MC could be established. The Aw value may be used as the criterion for honey industry, and then the MC of honeys can be calculated by the specific linear equation between Aw and MC.

Funding: This research received no external funding.

Acknowledgments: The author would like to thank the Ministry of Science and Technology of the Republic of China for financially supporting this research under Contract No. MOST-104-2313-B-005-031.

Conflicts of Interest: The author declare no conflict of interest.

References

1. Krell, R. *Value-Added Products from Beekeeping*; Food & Agriculture Organization of the United Nations: Rome, Italy, 2013.
2. Grégrová, A.; Kružík, V.; Vrácovská, E.; Rajchl, A.; Čížková, H. Evaluation of factors affecting crystallization of disparate set of multi-flower honey samples. *Agron. Res.* **2015**, *13*, 1215–1226.
3. Root, A.I.; Root, E.R. *The ABC and Xyz of Bee Culture*; Kessinger Publishing: Whitefish, MT, USA, 2010.
4. Rockland, L.B. Water activity: Theory and applications to food. In *IFT Basic Symposium Series*; Springer: New York, NY, USA, 1987.

5. Abramovic, H.; Jamnik, M.; Burkan, L.; Kač, M. Water activity and water content in Slovenian honeys. *Food Control.* **2008**, *19*, 1086–1090. [CrossRef]

6. Cavia, M.M.; Fernández-Muiño, M.A.; Huidobro, J.F.; Sancho, M.T. Correlation between moisture and water activity of honeys harvested in different years. *J. Food Sci.* **2004**, *69*, 368–370. [CrossRef]

7. Chirife, J.; Zamora, M.C.; Motto, A. The correlation between water activity and moisture in honey: Fundamental aspects and application to Argentine honeys. *J. Food Eng.* **2006**, *72*, 287–292. [CrossRef]

8. Gleiter, R.A.; Horn, H.; Isengard, H.D. Influence of type and state of crystallisation on the water activity of honey. *Food Chem.* **2006**, *96*, 441–445. [CrossRef]

9. Sanjuán, E.; Estupiñán, S.; Millán, R.; Castelo, M.; Penedo, J.C.; Cardona, A. Contribution to the quality evaluation and the water activity prediction of La Palma Island honey. *J. Food Qual.* **1997**, *20*, 225–234. [CrossRef]

10. Zamora, M.C.; Chirife, J. Determination of water activity change due to crystallization in honeys from Argentina. *Food Control* **2006**, *17*, 59–64. [CrossRef]

11. Zamora, M.C.; Chirife, J.; Roldan, D. On the nature of the relationship between water activity and % moisture in honey. *Food Control* **2006**, *17*, 642–647. [CrossRef]

12. Beuchat, L.R. Influence of water activity on growth, metabolic activities and survival of yeasts and molds. *J. Food Prot.* **1983**, *46*, 135–141. [CrossRef]

13. Beckh, G.; Wessel, P.; Lüllmann, C. Natü rliche best & teile des honigs: Hefen und deren stoffwechselprodukte—Teil 2: Der wassergehalt und die wasseraktivitä t als qualitätsparameter mit bezug zum hefewachstum. *Deutsche Lebensmittel-Rundschau* **2004**, *100*, 14–17.

14. Ruegg, M.; Blanc, B. The water activity of honey and related sugar solutions. *LWT-Food Sci.* **1981**, *14*, 1–6.

15. Alcalá, M.; Gómez, R. Cálculo, de la actividad de agua de la miel. *Aliment Equip Technolo.* **1990**, *9*, 99–100.

16. Estupiñan, S.; Sanjuán, E.; Millán, R.; González-Cortes, M.A.; Cálculo, Y. Aplicación de modelos de predicción de actividad de agua en mieles artesanales. *Microbial Aliments Nutr.* **1998**, *16*, 259–264.

17. Millán, R.; Tudela, L.; Estupiñan, S.; Castelo, M.; Sanjuán, E. Contribución al cálculo de la actividad de agua en miel: Modelo de predicción de actividad de agua en mieles de Las Palmas. *Alimentaria* **1995**, *268*, 77–79.

18. Pérez, A.; Sánchez, V.; Baeza, R.; Zamora, M.C.; Chirife, J. Literature review on linear regression equations for relating water activity to moisture content in floral honeys: Development of a weighted average equation. *Food Bioprocess Tech.* **2009**, *2*, 437–440. [CrossRef]

19. Salamanca, G.G.; Pérez, F.C.; Serra, B.J.A. Determinación de la Actividad de agua en mieles Colombianas de las zonas de Bocayá s Tolima. Apiservices—Galería Apícola Virtual. 2001. Available online: https://www.apiservices.biz/es/articulos/ordenar-por-popularidad/715-determinacion-de-la-actividad-de-agua-en-mieles-colombianos (accessed on 12 July 2018).

20. Kutner, M.H.; Nachtheim, C.J.; Neter, J. *Applied Linear Regression Models*, 4th ed.; McGraw-Hill Education: New York, NY, USA, 2004; pp. 294–335.

21. Myers, R.H. *Classical and Modern Regression with Applications*, 2nd ed.; Duxbury: Pacific Grove, CA, USA, 1998; pp. 135–153.

22. Weisberg, S. *Applied Linear Regression*, 4th ed.; Wiley: New York, NY, USA, 2013.

23. Chen, A.; Chen, H.; Chen, C. Use of temperature and humidity sensors to determine moisture content of Oolong tea. *Sensors* **2014**, *14*, 15593–15609. [CrossRef]

24. Chen, H.; Chen, C. Equilibrium relative humidity method used to determine the sorption isotherm of autoclaved aerated concrete. *Build Environ.* **2014**, *81*, 427–435. [CrossRef]

25. Chen, C. Sorption isotherms of sweet potato slices. *Biosyst. Engin.* **2002**, *83*, 85–95. [CrossRef]

26. Chen, C. Moisture sorption isotherms of pea seeds. *J. Food Eng.* **2002**, *58*, 45–51. [CrossRef]

27. Chen, C.; Weng, Y. Moisture Sorption isotherms of Oolong tea. *Food Biopr. Technol.* **2010**, *3*, 226–233. [CrossRef]

28. Association Official Analytical Chemists. *Official Methods of Analysis of AOAC International*, 21th ed.; AOAC Scientific Publications: Arlington, VA, USA, 2019; Volume II.

29. Bogdanov, S.; Martin, P.; Lullman, C. Harmonised methods of the European Honey Commission. *Apidologie* **1997**, *1*, 59.
30. Cano, C.B.; Felsner, M.L.; Matos, J.R.; Bruns, R.E.; Whatanabe, H.M.; Almeida-Muradian, L.B. Comparison of methods for determining moisture content of citrus and eucalyptus Brazilian honeys by refractometry. *J. Food Compost. Anal.* **2001**, *14*, 101–109. [CrossRef]

MDPI

St. Alban-Anlage 66

4052 Basel

Switzerland

Tel. +41 61 683 77 34

Fax +41 61 302 89 18

www.mdpi.com

Foods Editorial Office

E-mail: foods@mdpi.com

www.mdpi.com/journal/foods

Lightning Source UK Ltd.
Milton Keynes UK
UKHW051041151220
375215UK00003B/204